Methods in Molecular Biology

Series Editor
John M. Walker
School of Life and Medical Sciences
University of Hertfordshire
Hatfield, Hertfordshire, UK

For further volumes:
http://www.springer.com/series/7651

For over 35 years, biological scientists have come to rely on the research protocols and methodologies in the critically acclaimed *Methods in Molecular Biology* series. The series was the first to introduce the step-by-step protocols approach that has become the standard in all biomedical protocol publishing. Each protocol is provided in readily-reproducible step-by-step fashion, opening with an introductory overview, a list of the materials and reagents needed to complete the experiment, and followed by a detailed procedure that is supported with a helpful notes section offering tips and tricks of the trade as well as troubleshooting advice. These hallmark features were introduced by series editor Dr. John Walker and constitute the key ingredient in each and every volume of the *Methods in Molecular Biology* series. Tested and trusted, comprehensive and reliable, all protocols from the series are indexed in PubMed.

MERS Coronavirus

Methods and Protocols

Edited by

Rahul Vijay

Department of Microbiology and Immunology, University of Iowa, Iowa City, IA, USA

Humana Press

Editor
Rahul Vijay
Department of Microbiology
and Immunology
University of Iowa
Iowa City, IA, USA

ISSN 1064-3745 ISSN 1940-6029 (electronic)
Methods in Molecular Biology
ISBN 978-1-0716-0210-2 ISBN 978-1-0716-0211-9 (eBook)
https://doi.org/10.1007/978-1-0716-0211-9

This Humana imprint is published by the registered company Springer Science+Business Media, LLC, part of Springer Nature.
The registered company address is: 233 Spring Street, New York, NY 10013, U.S.A.

Preface

Since the emergence of SARS-CoV in 2002, the "prophecy" that a CoV outbreak is always "around the corner" was proven by the emergence of MERS-CoV in mid-2012. But unlike SARS-CoV, which was contained within an year of emergence (thanks to the efforts of medical professionals and researchers), MERS-CoV cases continue to occur after 7 years. According to WHO, since September of 2012, at least 2428 cases have been confirmed with 838 deaths spread over 27 countries. With this book, we consolidate the various techniques and methodologies that are being currently used in the study of MERS-CoV. Since MERS-CoV is a close cousin of SARS-CoV, the approaches presented here will have varying degrees of redundancy with those used for SARS studies. Given that there were two outbreaks within 10 years, one should be wary of another CoV outbreak in the near future. The chapters in this book should be considered as an up-to-date description of techniques that have been used in the study of CoVs, and will act as a useful reference if any such new outbreaks do appear.

This book has been divided into four parts: (1) *Evolution and Entry of MERS-Coronavirus*; (2) *Genetic Alteration and Structural Determination of MERS-Coronavirus Proteins*; (3) *Quantitation of Virus and Antiviral Factors*, and (4) *Mouse Models for MERS-Coronavirus*.

Emergence of new variants of coronaviruses with different host tropism warrants a thorough investigation of their evolution and acquired adaptability to these hosts. Thus, we begin the book with a chapter that details various methodologies to study evolutionary genetics of MERS-CoV under selection pressure. Understanding how the virus enters a host cell to initiate an infection is key to designing strategies to prevent it. Therefore, chapters describing various methodologies to identify MERS-CoV entry pathways and characterizing the key proteins employed by the virus to do so are also presented here.

An important aspect of studying viruses involves the ability to alter their genome by reverse genetics and to recover recombinant viruses with defined mutations. Such approaches will help in studying the functions of specific genes and their effects on virus survival and pathogenesis. These strategies will also aid in determining different checkpoints in the progression of virus growth and proliferation and developing therapeutics to prevent pathogenesis. For this purpose, a chapter detailing methodologies to genetically alter MERS-CoV is provided. Along the same lines, another chapter follows that describes protocols for deducing the crystal structure of an essential viral protein, helicase, that is indispensable for viral RNA transcription and replication.

With continuous emergence of coronaviruses, there is a critical need for diagnostic tools that could be employed in the field to successfully contain and prevent infection(s). To this end, two of the five chapters in Part III are dedicated to quantification of MERS-CoV viral loads using ELISA- and qRT-PCR-based techniques. To measure host immune parameters, methodologies to detect anti-MERS-CoV antibodies using microneutralization, pseudotyped viral particles, as well as ELISA-based methodologies are described.

Finally, no studies are complete without having reliable animal model systems to reproducibly replicate the disease and pathology observed in human cases. Such animal models are necessary to study the pathogenesis of the virus as well as the immune response to it. One bottleneck to achieving this goal was the inability of MERS-CoV to naturally infect mice, due to the lack of specific entry receptors. Part IV of this book begins with a chapter that describes how genetic engineering has been employed to render mice susceptible to MERS-CoV. This chapter is followed by another chapter that describes how the error-prone gene replication machinery of CoVs was utilized to generate mouse-adapted strain of MERS-CoV. Using such mouse models, one can ask what factors are produced and secreted by the host cell upon successful entry and productive infection by MERS-CoV. While detecting and assaying such molecules are relevant to understanding the host immune response, it is critical that these assays be performed under conditions in which the virus is completely inactivated to prevent laboratory-based transmission. A chapter describing an exhaustive protocol that simultaneously inactivates virus while retaining the quality of samples for downstream analysis is also provided. Vaccine and immunotherapeutic strategies rely on comprehensively understanding the host immune response stands following infection. Initial stages of MERS-CoV infection are associated with an exuberant inflammatory response that is both beneficial and damaging to the host. So we present two chapters, one focusing on evaluating the activation and inflammatory activity of immune cells in lungs and the other evaluating the histopathological changes following infection.

Books like these are a testimony to the selflessness of the scientific and medical community and the noble cause to which they are committed. The attempt to incorporate some of the relevant techniques utilized in MERS-CoV research into one single book wouldn't have succeeded without the willingness and concerted effort of investigators to take time from their busy schedules and contribute in a timely manner. We thank all the investigators who contributed both directly and indirectly to finally bringing this book to print.

Iowa City, IA, USA *Rahul Vijay*

Contents

Contributors

ANURODH S. AGRAWAL • *Departments of Microbiology and Immunology, University of Texas Medical Branch, Galveston, TX, USA*

SAWSAN S. AL-AMRI • *Vaccines and Immunotherapy Unit, King Fahd Medical Research Center, King Abdulaziz University, Jeddah, Saudi Arabia*

ABDULLAH ALGAISSI • *Departments of Microbiology and Immunology, University of Texas Medical Branch, Galveston, TX, USA; Department of Medical Laboratories Technology, College of Applied Medical Sciences, Jazan University, Jazan, Saudi Arabia*

NAIF KHALAF ALHARBI • *Department of Infectious Disease Research, King Abdullah International Medical Research Center, Riyadh, Saudi Arabia; King Saud bin Abdulaziz University for Health Sciences, Riyadh, Saudi Arabia*

ABDULRAHMAN ALMASAUD • *Department of Infectious Disease Research, King Abdullah International Medical Research Center, Riyadh, Saudi Arabia; King Saud bin Abdulaziz University for Health Sciences, Riyadh, Saudi Arabia*

AMANDA P. BECK • *Department of Pathology, Albert Einstein College of Medicine, Bronx, NY, USA*

KENT J. BLOODSWORTH • *Biological Sciences Division, Pacific Northwest National Laboratory, Richland, WA, USA*

RUDRAGOUDA CHANNAPPANAVAR • *Department of Acute and Tertiary Care, University of Tennessee Health Science Center, Memphis, TN, USA; Department of Microbiology, Immunology and Biochemistry, University of Tennessee Health Science Center, Memphis, TN, USA; Institute for the Study of Host-Pathogen Systems, University of Tennessee Health Science Center, Memphis, TN, USA*

ADAM S. COCKRELL • *Durham, NC, USA*

SHENG CUI • *NHC Key Laboratory of Systems Biology of Pathogens, Institute of Pathogen Biology, Chinese Academy of Medical Sciences & Peking Union Medical College, Beijing, P. R. China*

ANTHONY R. FEHR • *Department of Molecular Biosciences, University of Kansas, Lawrence, KS, USA*

JOSHUA FUNG • *Department of Microbiology, Li Ka Shing Faculty of Medicine, The University of Hong Kong, Pokfulam, Hong Kong*

TOM GALLAGHER • *Department of Microbiology and Immunology, Loyola University Chicago, Maywood, IL, USA*

GAUTAMI G. GALPALLI • *Department of Microbiology and Immunology, Loyola University Chicago, Maywood, IL, USA*

MICHAEL P. HANTAK • *Department of Microbiology and Immunology, Loyola University Chicago, Maywood, IL, USA*

WEI HAO • *NHC Key Laboratory of Systems Biology of Pathogens, Institute of Pathogen Biology, Chinese Academy of Medical Sciences & Peking Union Medical College, Beijing, P. R. China*

ANWAR M. HASHEM • *Department of Medical Microbiology and Parasitology, Faculty of Medicine, King Abdulaziz University, Jeddah, Saudi Arabia; Vaccines and Immunotherapy Unit, King Fahd Medical Research Center, King Abdulaziz University, Jeddah, Saudi Arabia*

YOUNG-MO KIM • *Biological Sciences Division, Pacific Northwest National Laboratory, Richland, WA, USA*

JENNIFER E. KYLE • *Biological Sciences Division, Pacific Northwest National Laboratory, Richland, WA, USA*

SUSANNA K. P. LAU • *Department of Microbiology, Li Ka Shing Faculty of Medicine, The University of Hong Kong, Pokfulam, Hong Kong; State Key Laboratory of Emerging Infectious Diseases, The University of Hong Kong, Pokfulam, Hong Kong; Collaborative Innovation Centre for Diagnosis and Treatment of Infectious Diseases, The University of Hong Kong, Pokfulam, Hong Kong*

SARAH R. LEIST • *Department of Epidemiology, University of North Carolina-Chapel Hill, Chapel Hill, NC, USA*

MICHAEL LETKO • *Laboratory of Virology, Division of Intramural Research, National Institute of Allergy and Infectious Diseases, Rocky Mountain Laboratories, National Institutes of Health, Hamilton, MT, USA*

KUN LI • *Department of Pediatrics, Pappajohn Biomedical Institute, University of Iowa, Iowa City, IA, USA*

PAUL B. MCCRAY JR • *Department of Pediatrics, Pappajohn Biomedical Institute, University of Iowa, Iowa City, IA, USA; Department of Microbiology, University of Iowa, Iowa City, IA, USA*

THOMAS O. METZ • *Biological Sciences Division, Pacific Northwest National Laboratory, Richland, WA, USA*

DAVID K. MEYERHOLZ • *Department of Pathology, University of Iowa Carver College of Medicine, Iowa City, IA, USA*

JEAN K. MILLET • *Department of Microbiology and Immunology, College of Veterinary Medicine, Cornell University, Ithaca, NY, USA; Virologie et Immunologie Moléculaires, INRA, Jouy-en-Josas, France*

RONALD J. MOORE • *Biological Sciences Division, Pacific Northwest National Laboratory, Richland, WA, USA*

VINCENT MUNSTER • *Laboratory of Virology, Division of Intramural Research, National Institute of Allergy and Infectious Diseases, Rocky Mountain Laboratories, National Institutes of Health, Hamilton, MT, USA*

ERNESTO S. NAKAYASU • *Biological Sciences Division, Pacific Northwest National Laboratory, Richland, WA, USA*

CARRIE D. NICORA • *Biological Sciences Division, Pacific Northwest National Laboratory, Richland, WA, USA*

STANLEY PERLMAN • *Department of Microbiology and Immunology, University of Iowa, Iowa City, IA, USA; Department of Pediatrics, University of Iowa, Iowa City, IA, USA; State Key Laboratory of Respiratory Disease, National Clinical Research Center for Respiratory Disease, Guangzhou Institute of Respiratory Health, The First Affiliated Hospital of Guangzhou Medical University, Guangzhou, China*

ENYA QING • *Department of Microbiology and Immunology, Loyola University Chicago, Maywood, IL, USA*

AMY C. SIMS • *Department of Epidemiology, University of North Carolina at Chapel Hill, Chapel Hill, NC, USA*

CHIEN-TE K. TSENG • *Departments of Microbiology and Immunology, University of Texas Medical Branch, Galveston, TX, USA; Center for Biodefense and Emerging Infectious Disease, University of Texas Medical Branch, Galveston, TX, USA*

XINQUAN WANG • *School of Life Sciences, Tsinghua University, Beijing, China*

GARY R. WHITTAKER • *Department of Microbiology and Immunology, College of Veterinary Medicine, Cornell University, Ithaca, NY, USA*
PATRICK C. Y. WOO • *Department of Microbiology, Li Ka Shing Faculty of Medicine, The University of Hong Kong, Pokfulam, Hong Kong; State Key Laboratory of Emerging Infectious Diseases, The University of Hong Kong, Pokfulam, Hong Kong; Collaborative Innovation Centre for Diagnosis and Treatment of Infectious Diseases, The University of Hong Kong, Pokfulam, Hong Kong*
SHUYUAN ZHANG • *School of Life Sciences, Tsinghua University, Beijing, China*
HAIXIA ZHOU • *School of Life Sciences, Tsinghua University, Beijing, China*

Part I

Evolution and Entry of MERS-Coronavirus

Chapter 1

Studying Evolutionary Adaptation of MERS-CoV

Michael Letko and Vincent Munster

Abstract

Forced viral adaptation is a powerful technique employed to study the ways viruses may overcome various selective pressures that reduce viral replication. Here, we describe methods for in vitro serial passaging of Middle East respiratory syndrome coronavirus (MERS-CoV) to select for mutations which increase replication on semi-permissive cell lines as described in Letko et al., Cell Rep 24, 1730–1737, 2018.

Key words MERS-CoV, Forced adaptation, Experimental evolution, Cell culture, Semi-permissive cell line, Host restriction, Species barrier

1 Introduction

RNA viruses are ideal model organisms to study evolutionary genetics under selection. This is due to their large population sizes and short generation times, which are characterized by rapid accumulation of mutations relative to other organisms. Given the error-prone nature of viral RNA-dependent RNA polymerases, viral replication leads to the formation of a quasispecies [1–3]. Rather than one virus producing identical progeny during replication, a population of viruses is produced, each differing from one another by nucleotide substitutions or deletions as a result of errors incorporated by the RNA polymerase. While the majority of these mutations will have neutral or negative effects on viral fitness, a small subset of these mutations may prove beneficial and enhance the ability for certain variants to replicate despite selective pressures of interest such as the host immune response or an antiviral drug. Forced adaptation experiments have been used to determine viral mutations that facilitate escape from drugs [4–6], monoclonal antibodies [7, 8], host restriction factors [9–11], and species variation in host receptors [12–14] and to elucidate various viral mechanisms of infection and replication [15–17].

Within the laboratory setting, the strength of selective pressure can be adjusted by increasing or decreasing the levels of the

Rahul Vijay (ed.), *MERS Coronavirus: Methods and Protocols*, Methods in Molecular Biology, vol. 2099,
https://doi.org/10.1007/978-1-0716-0211-9_1, © Springer Science+Business Media, LLC, part of Springer Nature 2020

restrictive factor, thus facilitating the rapid expansion of viral variants within the population of quasispecies that can overcome the applied selective pressure. The ideal environment is "semi-permissive"—allowing only low levels of wild-type virus replication. Below is the method employed to adapt MERS-CoV to a semi-permissive host receptor, *Desmodus rotundus* DPP4. The techniques described below could be applicable to a wide range of experiments to better understand the adaptive capacity of various coronaviruses under specific selective pressures.

2 Materials

2.1 Cell Culture

1. Semi-permissive cells: baby hamster kidney (BHK) cells which have been transduced to stably express *Desmodus rotundus* DPP4 (*dr*DPP4 [12]. Briefly, the coding sequence for *dr*DPP4 was cloned into a lentiviral expression cassette also encoding for mcherry-T2A-puromycin-N-acetyltransferase-P2A (System Biosciences) and used to generate lentiviral particles [9] (*see* **Note 1**). BHK cells were infected with lentiviral particles and then grown in DMEM containing puromycin at a final concentration of 1 ug/mL.
2. Cell culture media: Dulbecco's Modified Eagles Medium (DMEM), 10% fetal bovine serum, 1% L-glutamine, 1% penicillin and streptomycin, and 1 μg/mL puromycin.
3. Passaging culture media: Dulbecco's Modified Eagles Medium (DMEM), 2% fetal bovine serum, 1% L-glutamine, 1% penicillin and streptomycin, and 1 μg/mL puromycin.
4. Light microscope to check cell cultures for cytopathic effects.

2.2 Passaging Experiment

1. 6-Well cell-culture cluster plates.
2. MERS-CoV/EMC2012, passage 6. This virus stock was grown in-house and titered by standard endpoint titration on Vero cells [18].

2.3 Directed Sequencing of MERS-CoV Spike

1. Viral RNA extraction mini kit.
2. Superscript IV reverse transcriptase cDNA production kit.
3. iProof High-fidelity PCR kit.
4. Agarose gel purification kit.
5. MERS-CoV Spike receptor-binding domain sequencing primers (*see* Table 1).
6. Sequence analysis software capable of multiple sequence alignment and viewing chromatograms.

Table 1
Primers for sequencing MERS-CoV spike

Primer number	Primer sequence	Primer orientation
1	ATGATACACTCAGTGTTTCT	Forward
2	TAGAAGGCAGCCCAAGCTTTT	Reverse
3	TTACGTAACTGCACCTTTATG	Forward
4	CATTTCACCTGGAACAGAGC	Reverse
5	AGATTCTACATATGGCCCCCT	Forward
6	TTAGTGAACATGAACCTTATGCGGC	Reverse

3 Methods

3.1 Prepare Cells for Viral Passaging

1. Plan number of conditions. At least three replicates (well of semi-permissive cells) for each forced adaptation experiment should be performed in parallel. Critically, parental cells or a cell line stably expressing an irrelevant protein should be included to control for any nonspecific cell culture mutations.
2. Grow semi-permissive BHK cells to confluency in appropriate format. One 75 cm^2 flask should be sufficient to seed at least three 6-well cluster plates.
3. Wash, trypsinize, count, and seed BHK cell lines (parental controls and semi-permissive) in cell culture media (10% FBS) at a density of 1.5×10^5 cells/mL in a 2 mL volume in each well of 6-well plates (*see* **Note 2**).

3.2 Infect Cells

1. Twenty-four hours later, replace media on seeded cell lines with 2 mL of fresh passaging culture media (2% FBS).
2. Infect cells with MERS-CoV/EMC2012 at a final MOI of 0.01 (Fig. 1).

3.3 Prepare Cells for Subsequent Passage and Passage Virus

1. After 48 h postinfection, prepare new cell culture plates to passage virus. Follow initial seeding conditions and plate at a density of 1.5×10^5 cells/mL in a 2 mL volume in each well of 6-well plates.
2. Twenty-four hours after seeding the new cells (72 h postinfection of previous culture), replace media on seeded cell lines with 2 mL of fresh passaging culture media (2% FBS).
3. After 72 h postinfection for previous culture, take a 500 μL of supernatant sample from the infected culture and store for downstream viral sequencing. Store supernatants at −80 °C.
4. Check previously infected cells for emergence of cytopathic effects (cell death, rounding-up, and detachment from cell

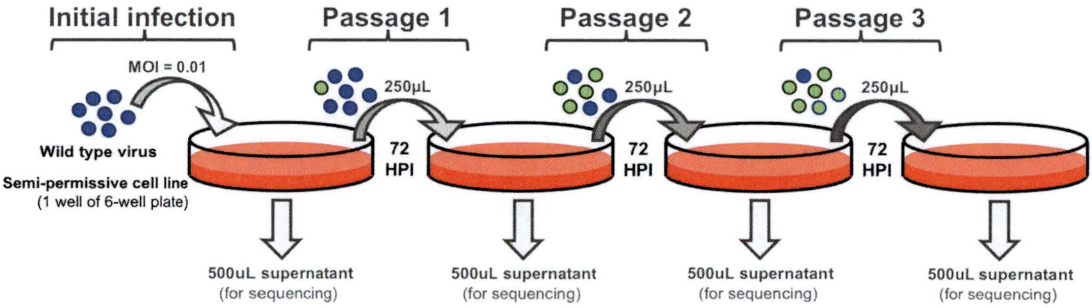

Fig. 1 Transduced cells are infected with wild-type stock. Approximately 72 hours later, supernatant from the infected cells is used to infect fresh cells as passage one. The process is repeated until the formation of cytopathic effects in culture. Supernatant from each passage is sequenced to detect the presence of adaptive mutations

culture plate in more than 50% of individual cultures) (*see* **Note 3**). If cytopathic effects are observed, this is strongly suggestive of viral adaptation to the semi-permissive cells. Proceed with step 3.4. Subsequent passages may be performed to select for further mutations that enhance viral replication in the semi-permissive cells (*see* **Note 4**).

5. If no cytopathic effects are observed, then begin next viral passage: from the previously infected culture, transfer 250 μL of supernatant to the new cell cultures seeded the day before.
6. Discard previously infected culture.
7. Repeat steps 1–6 until cytopathic effects are observed, indicative of viral adaptation.

3.4 Extract Viral RNA and Sequence Spike

1. Extract RNA from stored supernatants using the Qiagen viral RNA miniprep kit (Qiagen), following manufacturer's instructions.
2. Generate cDNA from extracted RNA using Superscript IV, following manufacturer's instructions.
3. Amplify select regions from viral cDNA using iProof high-fidelity PCR polymerase kit (Bio-Rad). Below are example PCR conditions for amplifying the MERS-CoV receptor-binding domain following the primer numbers listed in Subheading 2.2.5 of [12] (see Table 2).

31.5 μL	diH_2O
10 μL	iProof buffer
5 μL	dNTP mix
1 μL	forward primer (10 μM)
1 μL	reverse primer (10 μM)
0.5 μL	iProof enzyme
1 μL	cDNA (from Subheading 3.4, **step 2**)

Table 2
Primer pairs and expected product sizes for tiled MERS-CoV spike PCR amplification

Forward primer	Reverse primer	Expected PCR product size (bp)
1	2	940
3	4	1571
5	6	2447

PCR Cycling conditions

Temperature	Time	
98 °C	3 min	
98 °C 50 °C 72 °C	10 s 30 s 30 s	1.1.40 cycles
72 °C	5 min	
10 °C	Hold	

4. Gel purify PCR amplicons from 1% agarose using gel purification kit and following manufacturer's instructions.
5. Send each product for Sanger sequencing.
6. Check Sanger sequencing chromatograms for overlapping peaks, indicative of mutations within a mixed viral population, as further described in [12]

4 Notes

1. Importantly, this specific lentivector cassette is expressed under the Ef1α promoter, which allows for mid-level expression of the transgene as compared to other popular lentiviral transgene promoters such as CMV or CAGGS. This midlevel expression is ideal for semi-permissive selective pressure created by the transgene, in this case, *dr*DPP4.
2. The plating density of cells may vary from this suggested value, depending on growth kinetics. In general, cells should be plated to achieve approximately 80–90% confluency on the day of infection.
3. Cytopathic effects may be gradual to appear. To increase selective pressures on a viral population which is beginning to show signs of adaptation, one can apply a population bottleneck in the subsequent passage by reducing the amount of viral

supernatant passaged to the next cell culture. In this case, we recommend reducing the passage volume by approximately tenfold.

4. In our initial study [12], cytopathic effects were observed by the eighth passage; however, sequencing from earlier passages showed adaptive mutations emerging in the culture by the third passage. Depending on the strength of selection, the number of passages required to elicit adaptive mutations will vary.

Acknowledgment

Funding was provided by the Intramural research Program of the NIAID.

References

1. Andino R, Domingo E (2015) Viral quasispecies. Virology 479-480:46–51
2. Domingo E, Sheldon J, Perales C (2012) Viral quasispecies evolution. Microbiol Mol Biol Rev 76(2):159–216
3. Lauring AS, Andino R (2010) Quasispecies theory and the behavior of RNA viruses. PLoS Pathog 6(7):e1001005
4. Bazmi HZ et al (2000) In vitro selection of mutations in the human immunodeficiency virus type 1 reverse transcriptase that decrease susceptibility to (−)-beta-D-dioxolane-guanosine and suppress resistance to 3′-azido-3-′-deoxythymidine. Antimicrob Agents Chemother 44(7):1783–1788
5. Sato A et al (2006) In vitro selection of mutations in human immunodeficiency virus type 1 reverse transcriptase that confer resistance to capravirine, a novel nonnucleoside reverse transcriptase inhibitor. Antivir Res 70(2):66–74
6. McKimm-Breschkin JL et al (2012) In vitro passaging of a pandemic H1N1/09 virus selects for viruses with neuraminidase mutations conferring high-level resistance to oseltamivir and peramivir, but not to zanamivir. J Antimicrob Chemother 67(8):1874–1883
7. Shibata J et al (2007) Impact of V2 mutations on escape from a potent neutralizing anti-V3 monoclonal antibody during in vitro selection of a primary human immunodeficiency virus type 1 isolate. J Virol 81(8):3757–3768
8. Chai N et al (2016) Two escape mechanisms of influenza avirus to a broadly neutralizing stalk-binding antibody. PLoS Pathog 12(6): e1005702
9. Letko M et al (2015) Identification of the HIV-1 Vif and human APOBEC3G protein Interface. Cell Rep 13(9):1789–1799
10. Ooms M et al (2013) HIV-1 Vif adaptation to human APOBEC3H haplotypes. Cell Host Microbe 14(4):411–421
11. Ooms M, Letko M, Simon V (2017) The structural Interface between HIV-1 Vif and human APOBEC3H. J Virol 91(5):e02289-16
12. Letko M et al (2018) Adaptive evolution of MERS-CoV to species variation in DPP4. Cell Rep 24(7):1730–1737
13. Baric RS et al (1999) Persistent infection promotes cross-species transmissibility of mouse hepatitis virus. J Virol 73(1):638–649
14. McRoy WC, Baric RS (2008) Amino acid substitutions in the S2 subunit of mouse hepatitis virus variant V51 encode determinants of host range expansion. J Virol 82(3):1414–1424
15. Khasawneh AI et al (2013) Forced virus evolution reveals functional crosstalk between the disulfide bonded region and membrane proximal ectodomain region of HIV-1 gp41. Retrovirology 10:44
16. Das AT et al (2008) Optimization of the doxycycline-dependent simian immunodeficiency virus through in vitro evolution. Retrovirology 5:44
17. Westerhout EM et al (2005) HIV-1 can escape from RNA interference by evolving an alternative structure in its RNA genome. Nucleic Acids Res 33(2):796–804
18. van Doremalen N et al (2014) Host species restriction of Middle East respiratory syndrome coronavirus through its receptor, dipeptidyl peptidase 4. J Virol 88(16):9220–9232

Chapter 2

Evaluating MERS-CoV Entry Pathways

Enya Qing, Michael P. Hantak, Gautami G. Galpalli, and Tom Gallagher

Abstract

Middle East respiratory syndrome coronavirus (MERS-CoV) is an emerging zoonotic pathogen with a broad host range. The extent of MERS-CoV in nature can be traced to its adaptable cell entry steps. The virus can bind host-cell carbohydrates as well as proteinaceous receptors. Following receptor interaction, the virus can utilize diverse host proteases for cleavage activation of virus-host cell membrane fusion and subsequent genome delivery. The fusion and genome delivery steps can be completed at variable times and places, either at or near cell surfaces or deep within endosomes. Investigators focusing on the CoVs have developed several methodologies that effectively distinguish these different cell entry pathways. Here we describe these methods, highlighting virus-cell entry factors, entry inhibitors, and viral determinants that specify the cell entry routes. While the specific methods described herein were utilized to reveal MERS-CoV entry pathways, they are equally suited for other CoVs, as well as other protease-dependent viral species.

Key words Middle East respiratory syndrome (MERS), Coronavirus (CoV), Protease, Pseudovirus, Spike (S), Viral entry, Endosome, Virus concentration, Virus purification, Protease inhibitor, HR2 peptide, IFITM3, TMPRSS2, Transfection

1 Introduction

Middle East respiratory syndrome coronavirus (MERS-CoV) is endemic in bats, and also in dromedary camels, and can be transmitted zoonotically from camels to humans [1–6]. The virus was discovered in humans in 2012, and since then there has been over 2,000 laboratory-confirmed cases worldwide, with 35% of infected humans suffering fatal outcomes [7–9]. Although MERS-CoV zoonotic and human-to-human transmission rates have declined due to general awareness and improved hospital practices, there are continued possibilities for epidemics, and there is a need for preventive vaccines and therapeutic antivirals. Mechanistic insights into human MERS-CoV entry will promote vaccine and antiviral drug developments.

MERS-CoV, like all other coronaviruses, exists as enveloped extracellular particles with protruding spike (S) proteins. Infection is initiated through viral S protein binding to host cell receptors.

Rahul Vijay (ed.), *MERS Coronavirus: Methods and Protocols*, Methods in Molecular Biology, vol. 2099,
https://doi.org/10.1007/978-1-0716-0211-9_2,

Subsequent proteolytic cleavage of cell-bound S proteins triggers S protein-mediated coalescence (fusion) of viral and cell membranes. The triggering process involves a series of currently obscure S protein conformational changes, from "pre-fusion" metastable states to "extended fusion intermediates" to "post-fusion" collapsed states that exist after virus and cell membranes have coalesced and viral mRNA genomes have dispensed into the host cell cytoplasm [10, 11]. These events depend on cellular proteases, and as one may expect, the availability of particular proteases immediately following receptor engagement is a rate-limiting step in CoV entry.

Cellular proteases accumulate in distinct subcellular locations on the endocytic CoV entry pathway: serine proteases such as trypsin and elastase are extracellular; the type II transmembrane serine proteases (TTSPs) are anchored into plasma membranes; and the cysteine-type cathepsin proteases are enriched in endosomes [12, 13]. For several CoV infections, not all of these proteases are required; however, it is possible that each has distinct potential to activate fusion such that a productive infection ensues. There is evidence that particular CoVs have "preferred" in vivo entry routes (e.g., MERS-CoV and 229E-CoV prefer plasma membrane entry, while some MHV-CoVs prefer endosome entry [14–19], (*see* Fig. 1). Knowledge of these preferred routes, and their relation to virus-induced disease, is necessary to identify virus variants that might have high transmissibility and disease potential, and to recognize the host factors that might be targeted therapeutically such that infections are suppressed at the cell entry stage.

Here we provide protocols to dissect CoV entry pathways. These include procedures for pseudovirus production, particle purification and concentration, as well as specific assays to differentiate CoV entry pathways. While the protocols are set for characterizing MERS-CoV entry, they can be readily adjusted to evaluate other CoV and other protease-dependent virus entry events.

2 Materials

2.1 Particle Production

1. 150 mm Tissue culture dishes.
2. HEK-293T cells.
3. 293T cell media: Dulbecco's Modified Eagle Media (DMEM) with L-glut, 4.5 g/l glucose and 100 mM sodium pyruvate, additional supplements include 10% fetal bovine serum, 10 mM HEPES, 0.1 mM nonessential amino acids, 100 U/ml penicillin G, and 100 μg/ml streptomycin.
4. Transfection media: DMEM with L-glut, 4.5 g/l glucose and 100 mM sodium pyruvate, and 10% fetal bovine serum.

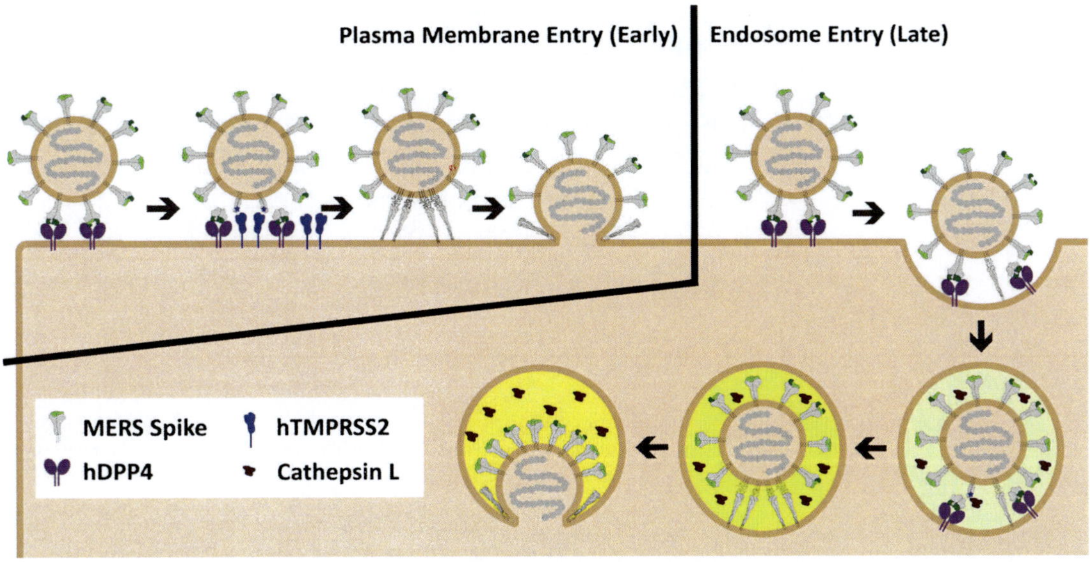

Fig. 1 MERS-CoV enters host either at or near the plasma membrane or in the endosomes. The MERS-CoV spike (S) proteins (gray) engage human DiPeptidyl Peptidase 4 (hDPP4, purple) via their receptor-binding domains (green). Receptor engagement exposes protease cleavage sites (blue stars) on S proteins. If cell surface proteases such as hTMPRSS2 (blue) are present, S proteins are cleaved and viral fusion occurs at or near the plasma membrane. If hTMPRSS2 or similar cell-surface proteases are not present, then MERS-CoV is endocytosed, and can be triggered by endosomal proteases such as cathepsin L (brown) to complete viral entry

5. Serum-free media: DMEM with L-glut, 4.5 g/l glucose and 100 mM sodium pyruvate, additional supplements include 10 mM HEPES, 0.1 mM nonessential amino acids, 100 U/ml penicillin G, and 100 μg/ml streptomycin.
6. Polyethylenimine (PEI) at 1 mg/ml dissolved in ddH2O.
7. OptiMEM reduced serum medium.
8. Expression plasmids for MERS-CoV-spike.
9. Expression plasmid for HIV core-Fluc (pNL4.3HIVluc).
10. Transducing particle: VSVΔG-Fluc pseudotyped with Junin virus (JUNV) GP.

2.2 Particle Purification and Concentration

1. Centrifuge: Eppendorf 5810 or equivalent.
2. Ultracentrifuge: Beckman Coulter's or equivalent. SW28 swinging-bucket rotor, buckets, and Ultra-Clear tubes.
3. Falcon 15 and 50 ml conical centrifuge tubes.
4. Sucrose solution: 20% sucrose (w/v) in serum-free media.

2.3 Characterizing Viral Entry Pathways

1. Falcon 6-well and 96-well cell culture plates.
2. 5x Cell Culture Lysis Reagent (CCLR): 125 mM Tris–HCl pH 7.8, 10 mM DTT, 10 mM 1,2-diaminocyclohexane-N,N,N′,N′-tetraacetic acid, 50% glycerol, 5% Triton X-100.

3. Firefly luciferase substrate: 1 mM D-luciferin, 3 mM ATP, 15 mM MgSO4·H2O, 30 mM HEPES [pH 7.8].
4. Protease inhibitor cocktail: 200 μM Camostat, 20 μM proprotein convertase inhibitor, 20 μM E64D in serum-free media.
5. Vehicle control: DMSO in serum-free media at equivalent levels to the protease inhibitor cocktail.
6. CoV fusion antagonists: CoV species-matching HR2 peptides.
7. Expression plasmids for: hTMPRSS2, hCD9, hIFITM3.

3 Methods

Carry out all incubations at 37 °C with 5% CO_2 unless otherwise specified.

3.1 VSV-Based Pseudovirus Production (see Note 1)

1. Plate enough 293T cells (5×10^6) in 20 ml into a 15 cm dish to reach 80% confluency on the next day.
2. On the following day, make transfection mixture by adding 20 μg of MERS-CoV-spike plasmid (*see* **Note 2**) and 110 μl of PEI into 2 ml of OptiMEM. Incubate the mixture in the dark for 15 min at room temperature.
3. Replace existing media with 20 ml of transfection media (pre-warmed to 37 °C, *see* **Note 3**). Add transfection mixture dropwise onto the cells. Incubate the cells for 6–8 h (*see* **Note 4**).
4. Replace transfection media with 20 ml of 293T cell media and incubate overnight.
5. Dilute 100× transducing particle (VSVΔG-Fluc pseudotyped with Junin virus (JUNV) GP, *see* **Note 5**) into 15 ml of pre-warmed serum-free media, which is then used to replace existing media on the transfected cells. Incubate cells for 2 h.
6. Remove supernatant, rinse cells with 10 ml of pre-warmed serum-free media three times, then add back 13 ml of pre-warmed 293T cell media. Incubate cells overnight.
7. Collect supernatant (first collection) with a 15 ml Falcon tube, add back 13 ml of pre-warmed 293T cell media, and incubate cells overnight (*see* **Note 6**).
8. Spin supernatant at 300 × *g* for 10 min at 4 °C.
9. Transfer supernatant into a fresh tube and spin at 3000 × *g* for 10 min at 4 °C. Discard pellet.
10. Transfer supernatant into a fresh tube and freeze it at −80 °C.

11. On the following day, repeat **steps 7–10** (second collection).
12. On the final day, collect supernatant (third collection), discard cells, repeat **steps 8–10**.

3.2 HIV-Based Pseudovirus Production

1. Plate enough 293T cells (5×10^6) in 20 ml into a 15 cm dish to reach 80% confluency on the next day.
2. On the following day, make transfection mixture by adding 10 μg of MERS-CoV-spike plasmid, 10 μg of HIV core-Fluc-expressing plasmid, and 110 μl of PEI into 2 ml of OptiMEM. Incubate the mixture in the dark for 15 min at room temperature.
3. Replace existing media with 20 ml of transfection media (pre-warmed to 37 °C). Add transfection mixture dropwise onto the cells. Incubate the cells for 6–8 h.
4. Replace transfection media with 20 ml of 293T cell media and incubate overnight.
5. Remove supernatant, and add back 13 ml of pre-warmed 293T cell media. Incubate cells overnight.
6. Collect supernatant (first collection) with a 15 ml Falcon tube, add back 13 ml of pre-warmed 293T cell media, and incubate cells overnight.
7. Spin supernatant at 300 × *g* for 10 min at 4 °C.
8. Transfer supernatant into a fresh tube and spin at 3000 × *g* for 10 min at 4 °C. Discard pellet.
9. Transfer supernatant into a fresh tube and freeze it at −80 °C.
10. On the following day, repeat **steps 6–9** (second collection).
11. On the final day, collect supernatant (third collection), discard cells, repeat **steps 7–9**.

3.3 Particle Purification and Concentration

We noted that pseudoviruses lose their transduction capabilities (up to 90%!) upon exposure to the high g-forces (~100,000 × *g*) commonly used in traditional viral concentration methods. Therefore, we adopted a low-speed viral concentration and purification protocol that achieves viral concentration without compromising viral transduction capabilities.

1. Thaw and pool collected pseudovirus-containing supernatants (*see* Subheadings 3.1 or 3.2).
2. Transfer 32 ml of the pooled supernatant into a SW28 Ultra-Clear tube.
3. Use a 3 ml syringe with needle to add a cushion of 3 ml of 20% sucrose to the bottom of the tube. Eject from syringe slowly to avoid sucrose mixing with the sample. After the placement of the cushion, gently add the remaining sample (~3 ml) into the

tube. If there is still space, add serum-free media to the brim of the Ultra-Clear tube.

4. Load the Ultra-Clear tube into a SW28 bucket, and spin at 6500 rpm (5591 × *g*) for 18 h at 4 °C.
5. After the spin, carefully take out the Ultra-Clear tube, and remove all the supernatant without disturbing the pellet at the bottom center (may be invisible). Quickly add back ~350 ml of serum-free media, and gently resuspend the pellet with a 1 ml pipette (*see* **Note 8**). Aliquot and store the fully resuspended sample (now 100× and purified) at −80 °C for future use.

3.4 Characterizing CoV Entry Pathways

Subsequent to receptor engagement, CoV spikes require proteolytic cleavage to trigger membrane fusion. Two classes of cellular proteases have been identified to trigger CoV fusions: Serine proteases that are either secreted or expressed on the plasma membrane [14, 17], and cysteine proteases that resides in the endosome [18, 19]. Therefore, the utilization of specific proteases by a given CoV also dictates its site of entry. We utilize several assays to differentiate the preferred site of entry for wild-type (WT) and mutant MERS spikes, the efficacies of entry inhibitors, and the identification of pro- or antiviral host factors.

3.4.1 Characterizing CoV Entry Kinetics Using Protease Inhibitor Cocktails (see Note 9)

The CoV site of entry is correlated with entry kinetics, with viral entry at the plasma membrane being "early," and entry through endosomes "late," in relation to the different virus trafficking times prior to membrane fusion [15–17]. Together with the knowledge that CoV entry requires proteolysis, we utilize a time-course assay to characterize CoV entry kinetics, where protease inhibitors are added at various time points to arrest future entry, allowing readouts for infection within short inoculation time windows.

1. Plate sufficient permissive cells into a 96-well plate, 60 μl per well, to reach 95% confluency on the next day.
2. On the following day, add 40 μl of MERS spike-bearing viral particles (Subheadings 3.1 or 3.2) onto cells. Incubate the plate at 4 °C for 1 h to allow viral binding.
3. After 1 h, remove unbound particles by aspirating the supernatant (*see* **Note 10**). Add back 50 μl of fresh media. Incubate the plate at 37 °C.
4. At various time intervals (0, 5, 10, 15, 20, 25, 30, 35, 40, 45, 60 min, and so on, *see* **Note 11**), add 50 μl of protease inhibitor cocktail to the cells. At time interval of 0 min, also have a condition where 50 μl of vehicle control (*see* **Note 12**) is added instead. Leave the drugs/vehicle on cells and incubate at 37 °C overnight (~18 h post viral inoculation).

5. Remove media, add 50 μl of 1×CCLR (5x CCLR diluted in ddH2O) to lyse cells. Freeze the plate at −80 °C for 30 min (*see* **Note 13**). Thaw the plate and transfer 20 μl to a white reading plate to analyze firefly luciferase activity. Normalize enzyme activity from all conditions to the vehicle control, which is set to "100%". Plot data as "% viral entry."

3.4.2 Characterizing Entry Routes Using Specific Protease Inhibitors

The entry assay (Subheading 3.4.1) can sensitively distinguish viruses with accelerated or delayed kinetics, but it is frequently time and reagent consuming. The assay can be simplified by an alternative where titration of a particular inhibitor is applied for a short period of time.

1. Plate sufficient permissive cells into a 96-well plate, 60 μl per well, to reach 95% confluency on the next day.
2. On the following day, at −1 h, remove media and add 60 μl of protease inhibitor or vehicle-containing serum-free media (1–1000 μM camostat to inhibit plasma membrane protease TMPRSS2, or 1–1000 μM E64D to inhibit endosomal cathepsins). Incubate cells for 1 h at 37 °C.
3. Add 40 μl of MERS spike-bearing viral particles (Subheading 3.1 or 3.2) onto cells. Incubate the plate at 37 °C for 2 h to allow viral entry.
4. After 2 h, remove unbound particles by aspirating the supernatant. Rinse twice with 100 μl PBS. Add back 50 μl of fresh media. Incubate the plate at 37 °C overnight (~18 h post viral inoculation).
5. Remove media, add 50 μl of 1×CCLR to lyse cells. Freeze the plate at −80 °C for 30 min. Thaw the plate and transfer 20 μl to a white reading plate to analyze firefly luciferase activity. Normalize enzyme activity from all conditions to the vehicle control, which is set to "100%." Plot data as "% viral entry."

3.4.3 Characterizing Entry Routes Using Spike Fusion Antagonists

The protocol described in Subheading 3.4.2 is flexible and can be tailored to evaluate different viral inhibitors. These include fusion inhibitors. CoV spike proteins facilitate membrane fusion by transiting from "extended fusion intermediates" to "post-fusion" conformations. This transition requires interactions between antiparallel helices termed heptad repeat region 1 (HR1) and 2 (HR2) [20, 21], and can be arrested by exogenous HR2 peptides, which bind to the fusion intermediates. Typically, HR2 peptides do not enter endosomes and therefore only arrest viruses that transition into intermediate conformations extracellularly, i.e., at target cell plasma membranes. However, lipid-conjugated HR2 peptides can bind plasma membranes and endocytose, accumulating in endosomes such that they will arrest intracellular virus-cell membrane fusion. Using a modified Subheading 3.4.2, we tested the efficacy of native vs. lipid-conjugated HR2 on blocking CoV endosomal entry [22].

1. Plate sufficient permissive cells into a 96-well plate, 60 μl per well, to reach 95% confluency on the next day.
2. On the following day, at −1 h, remove media and add 60 μl of native vs. lipid-conjugated HR2 peptides at 0.01–1 μM, or vehicle control. Incubate cells for 1 h at 37 °C.
3. Add 40 μl of MERS spike-bearing viral particles (Subheadings 3.1 or 3.2) onto cells. Incubate the plate at 37 °C for 1 h to allow viral entry.
4. After 1 h, remove unbound particles by aspirating the supernatant. Rinse twice with 100 μl PBS. Add back 50 μl of fresh media. Incubate the plate at 37 °C overnight (~18 h post viral inoculation).
5. Remove media, add 50 μl of 1x CCLR to lyse cells. Freeze the plate at −80 °C for 30 min. Thaw the plate and transfer 20 μl to a white reading plate to analyze firefly luciferase activity. Normalize enzyme activity from all conditions to the vehicle control, which is set to "100%." Plot data as "% viral entry."

3.4.4 Identifying Host Factors Participating in MERS-CoV Entry

With the ability to differentiate between MERS-CoV entry at the plasma membrane and within endosomes, we and others have identified several host factors that affect viral routes of entry [14, 15, 17, 23, 24]. These include but are not limited to: (1) transmembrane protease serine subtype 2 (hTMPRSS2), which facilitates MERS-CoV entry at the plasma membrane. (2) tetraspanin hCD9, which ferries the MERS-CoV receptor hDPP4 into close proximity with hTMPRSS2 to potentiate MERS-CoV entry at the plasma membrane. (3) interferon-induced transmembrane protein 3 (hIFITM3), which blocks CoV endosomal entry.

1. Plate sufficient permissive cells into a 6-well plate, 2 ml per well, to reach 85% confluency on the next day.
2. On the following day, transfect cells with expression plasmids for vector control, hTMPRSS2, hCD9, or hIFITM3 (*see* **Note 14**):
   (a) Make transfection mixture by adding 1 μg of plasmid and 12 μl of PEI into 200 μl of OptiMEM. Incubate the mixture in the dark for 15 min at room temperature.
   (b) Replace existing media with 2 ml of transfection media (pre-warmed to 37 °C). Add transfection mixture dropwise onto the cells. Incubate the cells for 6–8 h.
   (c) Replace transfection media with 2 ml of fresh media and incubate overnight.

3. On the next day, plate sufficient transfected cells into a 96-well plate, 60 μl per well, to reach 95% confluency on the next day.
4. Add 40 μl of MERS spike-bearing viral particles (Subheadings 3.1 or 3.2) onto cells. Incubate the plate at 4 °C for 1 h to allow viral binding.
5. After 1 h, remove unbound particles by aspirating the supernatant. Add back 50 μl of fresh media. Incubate the plate at 37 °C.
6. At various time intervals (0, 5, 10, 15, 20, 25, 30, 35, 40, 45, 60 min, and so on), add 50 μl of protease inhibitor cocktail to the cells. At time interval of 0 min, also have a condition where 50 μl of vehicle control is added instead. Leave the drugs/vehicle on cells and incubate at 37 °C overnight (~18 h post viral inoculation).
7. Remove media, add 50 μl of 1×CCLR (5×CCLR diluted in ddH2O) to lyse cells. Freeze the plate at −80 °C for 30 min. Thaw the plate and transfer 20 μl to a white reading plate to analyze firefly luciferase activity. Normalize enzyme activity from all conditions to the vehicle control, which is set to "100%." Plot data as "% viral entry."

4 Notes

1. The current protocol describes particle production in 15-cm diameter plates. This protocol can be scaled up or down as long as the ratio of all components remains constant. However, for consistency and reproducibility we do not recommend producing particles in container sizes smaller than 5-cm diameter.
2. When using PEI as a transfection reagent, use 1 μg of DNA per million cells.
3. 293T cells can detach from the plate, so exert care when replacing media. Always add back media after aspiration as soon as possible to prevent cell-drying. When adding media, liquid should land on the side wall of the plate, not directly onto the cells. Always pre-warm media.
4. The DNA:PEI ratio (1:~6) is optimized specifically for a transfection period of 6–8 h. If transfection is allowed to go overnight, lower the DNA:PEI ratio to 1:3.
5. JUNV GP-vsvΔG pseudovirus was chosen for its short half-life, which reduces the amount of inoculum JUNV GP pseudovirus contaminating the desired CoV-spike- vsvΔG.
6. This protocol maximizes pseudovirus production by taking three consecutive harvests from the producer cells (24–48,

48–72, and 72–96 h post-transfection). Each harvest period brings similar pseudovirus yields.

7. At the indicated spin speed, the k-factor is 4556. Since the sedimentation coefficients of VSV [25] or HIV [26] pseudovirus are around 500, a 9-h spin is sufficient. The actual spin time is doubled to insure that the pseudoviruses are pelleted through the more viscous 20% sucrose cushion.
8. When resuspending pellet, set 1 ml pipette to 200 μl and pipette up and down gently to avoid bubbles. Make sure to rinse the entire bottom of the tube to maximize collection.
9. The current protocol describes viral entry into cells seeded in 96-well plates. This protocol can be scaled up as long as the ratio of all components remains constant.
10. After removing media via aspiration, return fresh media to cells as soon as possible to prevent cell drying. Aspirate and return media from a maximum of 24 wells at a time.
11. CoV entry kinetics are unique to each virus and host cell combination. In LET-1 cells, MERS-CoV entry completes in around 1 h. In HeLa cells, 229E-CoV entry takes 4+ hours to complete. Pilot experiments with larger time intervals (30–60 min) are recommended.
12. The vehicle control for protease inhibitors is DMSO, which is cytotoxic at high concentrations. Pilot experiments are recommended to identify nontoxic DMSO concentrations.
13. Samples can be kept at −80 °C for up to a month before reading Fluc. To further prevent signal loss, protease inhibitor cocktail can be added to 1x CCLR before use.
14. For cell types that are killed by PEI:DNA transfection, lipid-based transfection vectors such as lipofectamine can be used instead. For cell types that are resistant to transfection, retro- or lenti-viral-based or adenovirus-based transduction systems can be used to introduce genes of interest [15].

Acknowledgment

This work is supported by NIAID grant AI060699 and by the Loyola University Chicago Center for Translational Research and Education.

References

1. Middle East respiratory syndrome coronavirus (MERS-CoV). https://www.who.int/en/news-room/fact-sheets/detail/middle-east-respiratory-syndrome-coronavirus-(mers-cov). Accessed 4 Mar 2019
2. Sabir JSM, Lam TTY, Ahmed MMM et al (2016) Co-circulation of three camel

coronavirus species and recombination of MERS-CoVs in Saudi Arabia. Science 351:81–84. https://doi.org/10.1126/science.aac8608

3. Reusken CBEM, Haagmans BL, Müller MA et al (2013) Middle East respiratory syndrome coronavirus neutralising serum antibodies in dromedary camels: a comparative serological study. Lancet Infect Dis 13:859–866. https://doi.org/10.1016/S1473-3099(13)70164-6
4. Haagmans BL, Al Dhahiry SHS, Reusken CBEM et al (2014) Middle East respiratory syndrome coronavirus in dromedary camels: an outbreak investigation. Lancet Infect Dis 14:140–145. https://doi.org/10.1016/S1473-3099(13)70690-X
5. Memish ZA, Mishra N, Olival KJ et al (2013) Middle East respiratory syndrome coronavirus in bats, Saudi Arabia. Emerg Infect Dis 19:1819–1823. https://doi.org/10.3201/eid1911.131172
6. Gilardi K, Byarugaba DK, Baric RS et al (2017) Further evidence for bats as the evolutionary source of Middle East respiratory syndrome coronavirus. MBio 8:1–13. https://doi.org/10.1128/mbio.00373-17
7. (2019)WHO | Middle East respiratory syndrome coronavirus (MERS-CoV). In: WHO. https://www.who.int/emergencies/mers-cov/en/. Accessed 4 Mar 2019
8. Fouchier RAM, van Boheemen S, Osterhaus ADME et al (2012) Isolation of a novel coronavirus from a man with pneumonia in Saudi Arabia. N Engl J Med 367:1814–1820. https://doi.org/10.1056/nejmoa1211721
9. Hijawi B, Abdallat M, Sayaydeh A et al (2013) Novel coronavirus infections in Jordan, April 2012: epidemiological findings from a retrospective investigation. East Mediterr Health J 19(Suppl 1):S12–S18
10. Li F (2016) Structure, function, and evolution of coronavirus spike proteins. Annu Rev Virol 3:237–261. https://doi.org/10.1146/annurev-virology-110615-042301
11. Hulswit RJG, de Haan CAM, Bosch B-J (2016) Coronavirus spike protein and tropism changes. Adv Virus Res 96:29–57. https://doi.org/10.1016/bs.aivir.2016.08.004
12. Millet JK, Whittaker GR (2015) Host cell proteases: critical determinants of coronavirus tropism and pathogenesis. Virus Res 202:120–134. https://doi.org/10.1016/j.virusres.2014.11.021
13. Bugge TH, Antalis TM, Wu Q (2009) Type II Transmembraneserine proteases. J Biol Chem 284:23177–23181. https://doi.org/10.1074/jbc.R109.021006
14. Park J, Li K, Barlan A et al (2016) Proteolytic processing of Middle East respiratory syndrome coronavirus spikes expands virus tropism. Proc Natl Acad Sci U S A 113:12262–12267. https://doi.org/10.1073/pnas.1608147113
15. Earnest JT, Hantak MP, Li K et al (2017) Thetetraspanin CD9 facilitates MERS-coronavirus entry by scaffolding host cell receptors and proteases. PLoS Pathog 13: e1006546. https://doi.org/10.1371/journal.ppat.1006546
16. Li K, Wohlford-Lenane CL, Channappanavar R et al (2017) Mouse-adapted MERS coronavirus causes lethal lung disease in human DPP4 knockin mice. Proc Natl Acad Sci 114: E3119–E3128. https://doi.org/10.1073/pnas.1619109114
17. Shirato K, Kanou K, Kawase M, Matsuyama S (2017) Clinical isolates of human coronavirus 229E bypass the endosome for cell entry. J Virol 91. https://doi.org/10.1128/JVI.01387-16
18. Qiu Z, Hingley ST, Simmons G et al (2006) Endosomal proteolysis by cathepsins is necessary for murine coronavirus mouse hepatitis virus type 2 spike-mediated entry. J Virol 80:5768–5776. https://doi.org/10.1128/JVI.00442-06
19. Matsuyama S, Taguchi F (2009) Two-step conformational changes in a coronavirus envelope glycoprotein mediated by receptor binding and proteolysis. J Virol 83:11133–11141. https://doi.org/10.1128/JVI.00959-09
20. Bosch BJ, Martina BEE, Van Der Zee R et al (2004) Severe acute respiratory syndrome coronavirus (SARS-CoV) infection inhibition using spike protein heptad repeat-derived peptides. Proc Natl Acad Sci U S A 101:8455–8460. https://doi.org/10.1073/pnas.0400576101
21. Bosch BJ, Rossen JWA, Bartelink W et al (2008) Coronavirus escape from heptad repeat 2 (HR2)-derived peptide entry inhibition as a result of mutations in the HR1 domain of the spike fusion protein. J Virol 82:2580–2585. https://doi.org/10.1128/JVI.02287-07
22. Park JE, Gallagher T (2017) Lipidation increases antiviral activities of coronavirus fusion-inhibiting peptides. Virology 511:9–18. https://doi.org/10.1016/j.virol.2017.07.033
23. Wrensch F, Winkler M, Pöhlmann S (2014) IFITM proteins inhibit entry driven by the MERS-coronavirus spike protein: evidence for cholesterol-independent mechanisms. Viruses 6:3683–3698. https://doi.org/10.3390/v6093683

24. Zhao X, Sehgal M, Hou Z, et al (2017) Identification of residues controlling restriction versus enhancing activities of IFITM proteins on entry of human coronaviruses. J Virol 92. https://doi.org/10.1128/JVI.01535-17
25. Bradish CJ, Brooksby JB, Dillon JF (1956) Biophysical studies of the virus system of vesicular stomatitis. J Gen Microbiol 14:290–314. https://doi.org/10.1099/00221287-14-2-290
26. Vogt V (1997) Retroviral Virions and Genomes. Cold Spring Harbor Laboratory Press

Chapter 3

Biochemical Characterization of Middle East Respiratory Syndrome Coronavirus Spike Protein Proteolytic Processing

Gary R. Whittaker and Jean K. Millet

Abstract

The coronavirus spike envelope glycoprotein is an essential viral component that mediates virus entry events. Biochemical assessment of the spike protein is critical for understanding structure–function relationships and the roles of the protein in the viral life cycle. Coronavirus spike proteins are typically proteolytically processed and activated by host cell enzymes such as trypsin-like proteases, cathepsins, or proprotein-convertases. Analysis of coronavirus spike proteins by western blot allows the visualization and assessment of proteolytic processing by endogenous or exogenous proteases. Here, we present a method based on western blot analysis to investigate spike protein proteolytic cleavage by transient transfection of HEK-293 T cells allowing expression of the spike protein of the highly pathogenic Middle East respiratory syndrome coronavirus in the presence or absence of a cellular trypsin-like transmembrane serine protease, matriptase. Such analysis enables the characterization of cleavage patterns produced by a host protease on a coronavirus spike glycoprotein.

Key words Coronavirus, Spike protein, Virus entry, Middle East respiratory syndrome (MERS), Proteolytic processing, Host cell protease, Matriptase, Western blot, Transient transfection

1 Introduction

The coronavirus spike protein is a type I transmembrane protein that assembles as a trimer and projects outward from viral particles forming the distinctive corona or crown-like appearance of coronaviruses [1]. The spike protein controls to a large extent virus entry events as it accomplishes the critical functions of host cell receptor attachment and membrane fusion [2]. Coronavirus spike envelope glycoproteins are classified as class I viral fusion proteins [3]. Spike monomers are organized into a receptor-binding domain, named S1, and the fusion machinery domain, called S2 [4]. Many coronavirus spike proteins are proteolytically processed by host cell proteases at the junction between the S1 and S2 domains, at a site named S1/S2 (Fig. 1) [5]. An additional cleavage event occurs at a site found within the S2 domain, immediately

Rahul Vijay (ed.), *MERS Coronavirus: Methods and Protocols*, Methods in Molecular Biology, vol. 2099,
https://doi.org/10.1007/978-1-0716-0211-9_3, © Springer Science+Business Media, LLC, part of Springer Nature 2020

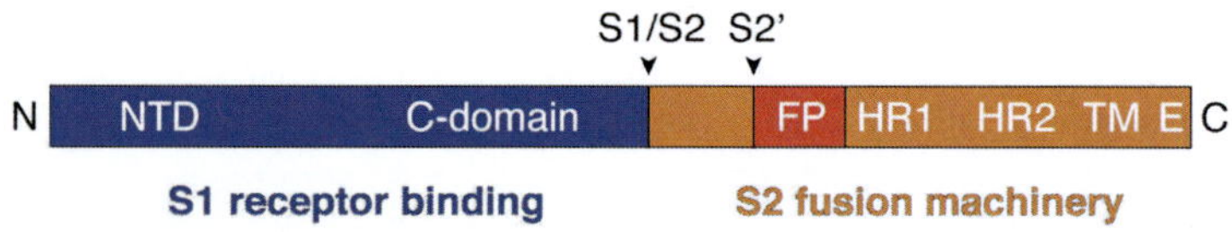

Fig. 1 Schematic of the coronavirus spike protein. Shown in the linear diagram are the S1 receptor binding domain and the S2 fusion machinery domain. S1/S2 denotes the proteolytic cleavage site between S1 and S2 domains. S2' is the cleavage site located within the S2 fusion domain, immediately upstream of the fusion peptide (FP). HR1: heptad repeat 1 region; HR2: heptad repeat 2 region; TM: transmembrane domain; E: endodomain

upstream of the fusion peptide, and termed S2' (Fig. 1) [6]. For coronaviruses, entry and viral fusion can occur at the plasma membrane or in endosomes, depending on protease availability and other microenvironmental cues [7, 8]. The cleavage event at S2' is analogous to the proteolytic processing of the prototypical class I viral fusion protein influenza virus hemagglutinin (HA) [9, 10]. Influenza virus entry occurs in the endosomal compartment. The critical HA proteolytic activation step releases the fusion peptide, located at the N-terminal tip of the fusion domain (HA2). Within maturing endosomes, the acidic pH triggers major conformational changes that allow membrane insertion of the fusion peptide and initiation of the merging of viral and host cell membranes [7].

Biochemical characterization of the spike activation step in the coronavirus life cycle is crucial to understand how this essential event impacts entry into host cells and modulates host cell and tissue tropism, host range and pathogenicity. Slight modifications of viral envelope glycoprotein cleavage sites via mutations can have a profound impact on host protease substrate recognition and activation, leading to changes in pathogenicity. This is typified by highly pathogenic strains of avian influenza (H5N1 strains) where modifications of the HA cleavage motif from a monobasic to a polybasic site switches the proteolytic activation mechanism [11]. Such cleavage site modification is accompanied by a change in host protease requirement from a trypsin-like protease to a more ubiquitously expressed furin-like protease, allowing for systemic infection and increased disease severity. Interestingly, coronavirus spike proteins and in particular the envelope glycoprotein of Middle East respiratory syndrome coronavirus (MERS-CoV) have been shown in cell culture to be cleaved by a number of host cell proteases such as cathepsins [12–14], trypsin-like proteases, notably membrane-bound transmembrane protease, serine 2 (TMPRSS2) [12–15], and members of the pro-protein convertase family of enzymes such as furin [16, 17]. The latter group of proteases may not be major players for typical routes of virus entry in respiratory epithelial cells [18], but could nonetheless be important activators for extra-pulmonary spread of MERS-CoV [17]. While various proteases have been shown to activate MERS-CoV spike in cell culture systems, it has recently been proposed that TMPRSS2 plays a predominant activating role in virus entry in respiratory epithelial cells [19].

Western blot analyses of spike protein cleavage products produced by host proteases is an important investigative tool allowing a better understanding of the roles of cleavage activation and proteases in the life cycle of coronaviruses. The aim of this chapter is to present a detailed protocol allowing the characterization by western blot of the proteolytic processing of the MERS-CoV spike protein, by transient co-transfection of a plasmid encoding MERS-CoV spike and increasing amounts of a plasmid encoding a host protease (*see* **Note 1**). An advantage of this transient transfection-based method compared to MERS-CoV infection-based experiments is that it does not require a biosafety level 3 facility to be performed. We will explain some possible difficulties of the method as well as provide tips and troubleshooting guides. The protease chosen to illustrate the method is matriptase or suppression of tumorigenicity 14 (ST-14) [20], an enzyme belonging to the type II transmembrane serine protease (TTSP) family, which also groups members such as human airway trypsin-like protease (HAT), TMPRSS2, and corin. The activity of matriptase is tightly regulated by the binding of hepatocyte growth factor activator inhibitor type 1 (HAI-1) [21], a Kunitz-type transmembrane serine protease inhibitor. Matriptase is broadly expressed by epithelial cells and throughout the respiratory tract, in particular in airway epithelial cells [22]. While it is synthesized as a membrane-bound protease, the catalytic domain of matriptase can be shed and act extracellularly. Importantly, while matriptase was found to activate the envelope glycoprotein of other viruses such as influenza virus HA [22, 23], its role in proteolytic processing of coronavirus spike protein has not been characterized. The analysis illustrated here allows to characterize the cleavage pattern induced by co-expression of MERS-CoV spike and matriptase in transiently transfected cells. In the absence of matriptase expression, the assay confirms processing of MERS-CoV spike at the S1/S2 site by an endogenous protease, likely furin or a furin-like enzyme as demonstrated previously [17]. Intriguingly, further cleavage, mediated by matriptase expression releases an as-yet uncharacterized spike fragment migrating at 25 kDa.

2 Materials

All cell culture materials should be kept sterile and manipulated within a biosafety cabinet. When not in use they should be stored at 4 °C. All liquid and solid waste materials should be discarded and/or properly inactivated in appropriate disposable waste containers. Solutions diluted in water should be prepared with ultrapurified water with a resistivity of 18.2 MΩ·cm at 25 °C.

2.1 Plasmids and Antibodies

1. pcDNA3.1-OPT-MERS-wt-S-C9. This plasmid encodes a full-length, wild-type (wt), mammalian codon-optimized sequence of the MERS-CoV spike gene from the EMC/2012 strain fused with a C9 bovine rhodopsin epitope tag at the C-terminus.
2. pcDNA3.1-hMatriptase. This plasmid contains the coding sequence of the human matriptase gene.
3. pcDNA3.1. This plasmid is used as an empty vector control plasmid.
4. Rabbit polyclonal antibody against MERS-CoV strain EMC/2012 spike protein.
5. Mouse monoclonal antibody (IgG_1) against the extracellular domain of human matriptase (clone D-7).
6. Horseradish peroxidase (HRP)-conjugated goat anti-rabbit IgG antibodies.
7. HRP-conjugated goat anti-mouse IgG antibodies.

2.2 Cell Culture Reagents and Materials

1. Dulbecco's phosphate buffered saline (DPBS) with calcium and magnesium.
2. Dulbecco's Modified Eagle Medium (DMEM).
3. Heat-inactivated fetal calf serum (FCS).
4. 1 M N-2-hydroxyethylpiperazine-N′-2-ethanesulphonic acid (HEPES).
5. 100× penicillin-streptomycin (PS) solution.
6. Human embryonic kidney (HEK) HEK-293 T/17 cells were obtained from the American Type Culture Collection. The/17 numbering refers to a clone that has been specifically selected to obtain higher transfection efficiencies. Cells were cultured in a 37 °C, 5% CO_2 incubator in Dulbecco's Modified Eagle Medium (DMEM) supplemented with 10% (vol/vol) FCS, 10 mM HEPES, 100 IU/mL penicillin, and 100 μg/mL streptomycin. For long-term storage, the cells can be frozen and stored in liquid nitrogen.
7. 1× Trypsin solution. 0.25% trypsin, 2.21 mM ethylenediaminetetraacetic acid (EDTA).
8. Cell counting slide with 10 counting grids.
9. Gibco™ Opti-minimal essential medium (Opti-MEM™) reduced serum medium (for transfections).
10. Cell culture flasks (75 cm^2) with vented caps.
11. 6-Well plate culture vessel.

2.3 Transfection and Cell Lysis Reagents

1. Lipofectamine® 2000 transfection reagent.
2. 10× Radioimmunoprecipitation assay (RIPA) buffer. 0.5 M Tris–HCl, 1.5 M NaCl, 2.5% deoxycholic acid, 10% NP-40, 10 mM EDTA, pH 7.4.
3. Protease inhibitor cocktail.
4. Sterile cell scrapers.
5. Ice bucket with fresh ice.
6. Rocking device.
7. Benchtop centrifuge for microcentrifuge tubes.

2.4 Protein Gel Migration Reagents and Materials

1. Electrophoresis System with appropriate power adaptor, or equivalent.
2. 4× Lithium dodecyl sulfate (LDS) sample loading buffer. 106 mM tris(hydroxymethyl)aminomethane (Tris)–HCl, 141 mM Tris Base, 2% LDS, 10% glycerol, 0.51 mM EDTA, 0.22 mM SERVA blue G250, 0.175 mM phenol red.
3. 10× Reducing reagent. 500 mM dithiothreitol (DTT).
4. PageRuler™ Plus Prestained Protein Ladder.
5. NuPAGE® 4–12% gradient Bis-Tris polyacrylamide pre-cast protein gels.
6. 20× NuPAGE® 3-(N-morpholino)propanesulfonic acid (MOPS) sodium dodecyl sulfate (SDS) running buffer. 50 mM MOPS, 50 mM Tris Base, 0.1% SDS, 1 mM EDTA, pH 7.7.

2.5 Transfer Reagents and Materials

1. Electrophoretic Transfer Cell with gel holder cassettes and with appropriate power adaptor, or equivalent.
2. Polyvinylidene fluoride (PVDF) blotting membrane.
3. Whatman® cellulose filter paper cutouts to the size of the area of gel to transfer. For one transfer, prepare 2 sets of 3 Whatman cutouts.
4. Transfer fiber pads. Their size should fit the gel holder cassette and should be similar to that of the Whatman paper cutouts. For 1 transfer, 1 set of 2 pads are needed.
5. 20× NuPAGE™ Transfer Buffer. 25 mM Bicine, 25 mM Bis-Tris (free base), 1 mM EDTA, pH 7.2.
6. Methanol.
7. Frozen ice pack to maintain cool temperatures during transfer.
8. Ice bucket with fresh ice.

2.6 Immunoblotting Reagents and Materials

1. Tris-buffered saline (TBS). 50 mM Tri–HCl, 150 mM NaCl, pH 7.5.
2. TBS-Tween-20 (TBS-T). TBS with addition of 0.05% Tween-20.
3. TBS-T-bovine serum albumin 2% (TBS-T-BSA2). TBS-T with addition of 2% (2 g for 100 mL) bovine serum albumin (for blocking step).
4. TBS-T-bovine serum albumin 1% (TBS-T-BSA1). TBS-T with addition of 1% (1 g for 100 mL) bovine serum albumin (for antibody incubations steps).
5. Small container for membrane incubations.

2.7 Chemiluminescence Western Blot Imaging

1. Enhanced chemiluminescence (ECL) reagents kit.
2. Chemiluminescence imager.

3 Methods

Perform steps involving opened sterile solutions or cell culture vessels under sterile conditions using a biosafety cabinet.

3.1 Cell Seeding

1. Observe under a light microscope the HEK-293 T/17 cells cultured in a 75 cm^2 flask and check that they are near-confluent, at around 80–90% (*see* **Note 2**).
2. Aspirate spent cell medium and wash cells once with 5 mL of pre-warmed (37 °C) DPBS.
3. Aspirate supernatant.
4. Trypsinize cells by adding 1 mL of trypsin solution directly to cells.
5. Place flask at a 37 °C 5% CO_2 incubator for 2–3 min (*see* **Note 2**).
6. Place flask back under biosafety cabinet and add 4 mL of pre-warmed (37 °C) FCS-containing DMEM.
7. Perform repetitive up-down pipetting to dissociate cells thoroughly.
8. Transfer dissociated cells to a sterile 50 mL conical tube.
9. Add 5 mL of pre-warmed (37 °C) FCS-containing DMEM to cells and vortex tube thoroughly.
10. Transfer 10 μL of the cell solution to a chamber of a cell counting slide.
11. Count cells under a light microscope.

Table 1
Quantities of plasmid DNA and Opti-MEM for transfection of one well of a 6-well plate

Reagent	Quantity for 1 well of a 6-well plate
pcDNA3.1-OPT-MERS-wt-S-C9	500 ng
pcDNA3.1-hMatriptase/pcDNA3.1-empty vector	0, 1, 2, 5, 10, 20, 50,100, and 200∗ ng
Opti-MEM	To 50 μL

Increase amounts proportionally to the number of replicate wells to transfect. We recommend adding an extra "safety" well in calculations to mitigate pipetting errors. For the different conditions tested in this protocol, the spike protein encoding plasmid quantity remains constant at 500 ng while the quantities of human matriptase encoding plasmid gradually increases from 0 to 200 ng. The asterisk (∗) denotes that for each matriptase encoding plasmid quantity a reciprocal amount of pcDNA3.1-empty vector should be added so that the total amount plasmid DNA remains constant at 200 ng (e.g., for the 10 ng matriptase encoding plasmid condition, there should be 200–10 = 190 ng of pcDNA3.1 empty vector added)

12. Dilute cells using pre-warmed (37 °C) FCS-containing DMEM to a concentration of 5×10^5 cells/mL.
13. Seed cells by adding 2 mL of diluted cell solution to each well of a 6-well plate.
14. Perform repetitive side-to-side, front-and-back plate movements on a horizontal surface, and gently tap on sides to evenly distribute cells (*see* **Note 3**).
15. Incubate cells overnight in a 37 °C 5% CO_2 incubator.

3.2 Transfection and Expression of MERS-CoV Spike and Matriptase

1. Observe HEK-293 T/17 cells in 6-well plate under a light microscope and check that they are approximately 70–80% confluent.
2. Dilute MERS-CoV S-encoding and human matriptase-encoding plasmid preparations with Opti-MEM reduced serum medium according to volumes shown on Table 1. For this assay, in each transfected well, the MERS-CoV spike plasmid amount remains constant (500 ng of plasmid DNA) while the quantities of plasmid encoding human matriptase is increased (0–200 ng of plasmid DNA). Include a condition using mock-transfected control (pcDNA3.1 empty vector control).
3. Dilute Lipofectamine 2000 solution with Opti-MEM reduced serum medium according to volumes shown on Table 2 for one well of a 6-well plate. Increase amounts proportionally to the number of wells to transfect and include extra "safety" wells to mitigate pipetting errors. Always add Lipofectamine 2000 to Opti-MEM solution and not the other way around (*see* **Note 4**).
4. Incubate plasmid DNA-Opti-MEM and Lipofectamine 2000-Opti-MEM solutions at room temperature for 5 min.

Table 2
Quantities of Lipofectamine™ 2000 transfection reagent and Opti-MEM for transfection of one well of a 6-well plate

Reagent	Quantity for 1 well of a 6-well plate
Lipofectamine™ 2000	3 μL
Opti-MEM	47 μL

Make a master mix by increasing proportionally the amounts of reagents to the numbers of wells to transfect and add additional "safety" wells in calculations to mitigate pipetting errors

5. Add Lipofectamine 2000-Opti-MEM solution to plasmid DNA-Opti-MEM solutions at a 1:1 vol/vol ratio (e.g., for one well of a 6-well plate add 50 μL of Lipofectamine 2000-Opti-MEM solution to 50 μL of plasmid DNA-Opti-MEM solution).
6. Incubate transfection solutions at room temperature for 20 min.
7. Aspirate spent medium of each cell culture well.
8. Add 1 mL of pre-warmed (37 °C) Opti-MEM to each well.
9. Add 100 μL per well of transfection solutions in a dropwise manner.
10. Rock plate gently.
11. Incubate plate in a 37 °C, 5% CO_2 incubator for 4–6 h.
12. Add 1 mL per well of FCS-containing DMEM without antibiotics (*see* **Note 5**).
13. Incubate plate in a 37 °C, 5% CO_2 incubator for 24 h.

3.3 Cell Lysis and Preparation of Protein Samples

To preserve protein sample integrity, all steps should be performed on ice (tubes and plates). Prechill all buffers/solutions on ice until the sample boiling step. The temperature of the microcentrifuge should also be set at 4 °C.

1. Prepare 50 mL of 1× RIPA buffer by diluting stock solution with ultrapure water.
2. Add 1 tablet of protease inhibitor cocktail (PIC) to the RIPA buffer.
3. Place tube on rocker until tablet dissolves completely.
4. Chill tube of RIPA buffer on ice.
5. Place plate of transfected cells on ice.
6. Aspirate gently the spent media of cells.
7. Wash cells gently by adding 1 mL per well of prechilled DPBS, avoiding pipetting directly onto cells.
8. Aspirate supernatants.

Table 3
Quantities of reagents to add to 300 μL of lysate sample

Reagent	Quantity for each sample
Protein sample	300 μL
4× LDS	115 μL
10× DTT	46 μL

We suggest to first prepare a master mix composed of proportionally increased amounts of 4× LDS and 10× DTT corresponding to the number of samples being prepared. As mentioned previously, include an extra "safety" sample in calculations to mitigate pipetting errors, and then add 161 μL of the mix to each sample of 300 μL

9. Lyse cells by adding 300 μL per well of prechilled RIPA buffer supplemented with PIC.
10. Incubate cells on ice with rocking for 10 min.
11. Use sterile scrapers for each well to completely detach lysed cells.
12. Transfer each cell lysate to a chilled microcentrifuge tube placed on ice.
13. Centrifuge samples at 15,000 × *g* on a benchtop microcentrifuge set at 4 °C for 20 min.
14. Transfer supernatants of each sample to a new set of chilled microcentrifuge tubes placed on ice. The pellets can be discarded.
15. Prepare LDS loading buffer and DTT reducing agent solution (Table 3) (*see* **Note 6**).
16. Add 161 μL of LDS-DTT solution to each sample.
17. Heat samples at 95 °C for 5 min (*see* **Note 6**).
18. Place tubes on ice to cool down for 1 min.
19. Perform a quick microcentrifugation step to spin down evaporated water on microcentrifuge tube caps.
20. Store samples at −20 °C (*see* **Note 7**).

3.4 Polyacrylamide Gel Electrophoresis

1. Make 1 L of 1× Bis-Tris gel running buffer by diluting buffer stock solution in ultrapure water.
2. Prepare pre-cast gel (Bis-Tris 4–12% gradient) by removing the comb and adhesive tape, rinse exterior casing with ultrapure water, and gently wash each lane well with ultrapure water (*see* **Note 8**).
3. Assemble pre-cast gel in the electrophoresis tank following the manufacturer's guidelines.

4. Add 1× Bis-Tris running buffer in the electrophoresis tank making sure that the pre-cast gel assembly is properly sealed and does not leak out into the outer parts of the tank.
5. Load 25 μL of each sample in individual gel lane wells and include a lane with protein ladder (10 μL).
6. Connect electrophoresis tank to power supply generator. Turn on power with constant voltage initially set at 100 V (*see* **Note 9**).
7. Check that protein samples are migrating downward by looking at migration front.
8. Incrementally increase voltage up to 200 V, within a 10–15 min timeframe (*see* **Note 9**).
9. Migrate samples until migration front reaches bottom of gel (approximately where the adhesive tape was located). Migration time typically lasts for a little over an hour.
10. Turn off power supply and remove gel from electrophoresis tank.

3.5 Electrophoretic Transfer of Protein Samples

1. Prepare 1 L of 1× transfer buffer with methanol (10% final) by diluting the buffer stock solution with ultrapure water (*see* **Note 10**).
2. Prechill transfer buffer on ice.
3. Incubate PVDF membrane cutouts (the size should cover the area of gel to transfer) in pure methanol for 10 min (*see* **Note 11**).
4. Discard methanol from membrane and immediately replace with transfer buffer.
5. Soak Whatman paper (6 paper cutouts per transfer) and fiber pads (2 pads per transfer) in transfer buffer.
6. De-cast carefully the polyacrylamide gel delicately and immediately place it in a container with transfer buffer (*see* **Note 12**).
7. Layer transfer components within a transfer cassette according to diagram shown in Fig. 2.
8. Roll out bubbles after layering PVDF membrane on gel using a clean serological pipette that has been humidified with transfer buffer (*see* **Note 12**).
9. Lock transfer components within transfer cassette.
10. Place transfer cassette in transfer electrophoretic tank being mindful of the direction of the electric current in the tank. In the transfer tank used here the black panel of the transfer cassette should directly face the black wall of the electrodes assembly.
11. Place frozen ice pack in transfer tank.

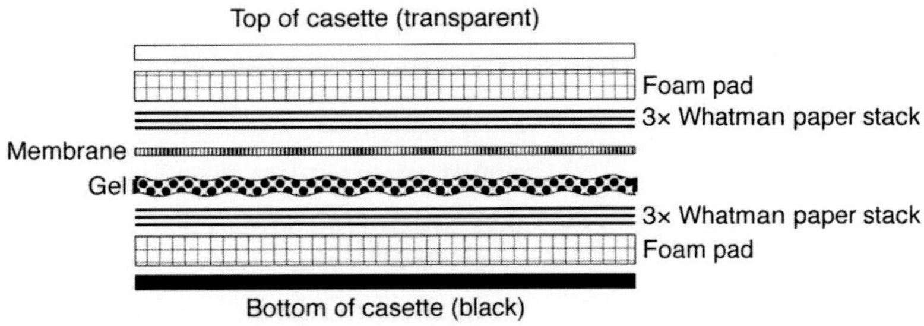

Fig. 2 Side-view diagram of transfer component stack within transfer cassette. Displayed in expanded view are the various components of the transfer stack to place in the transfer cassette. The transfer stack should be prepared in a container filled with transfer buffer. The exaggerated gaps between the different components shown here are for clarity only. In the actual transfer stack there should be no gaps or air bubbles between the different layers

12. Add chilled transfer buffer to transfer tank.
13. Place transfer tank in ice bucket containing fresh ice (*see* **Note 13**).
14. Connect transfer tank with power supply generator.
15. Turn power on using constant current set at 200 mA for 3 h (*see* **Note 13**).
16. Turn off power supply generator.

3.6 Immunoblotting

1. Prepare 1 L of TBS-T, 50 mL of TBS-T-BSA2 and 50 mL of TBS-T-BSA1 solutions (*see* **Note 14**).
2. Disassemble layered transfer components from transfer cassette avoiding dehydrating PVDF membrane.
3. Place PVDF membrane in TBS-T buffer in a small container immediately after disassembly of transfer components.
4. Replace TBS-T with TBS-T-BSA2 blocking buffer (*see* **Note 14**).
5. Incubate membrane for 1 h at room temperature with gentle rocking.
6. Prepare primary antibody dilutions in TBS-T-BSA1 dilution buffer (1/2000 for anti-MERS-CoV-S antibodies and 1/1000 for anti-matriptase antibodies) (*see* **Note 15**). Typically, 8–10 mL of antibody solution is enough to cover surface of a membrane in "mini-gel" format.
7. Incubate membrane with antibodies overnight at 4 °C with rocking.
8. Wash the membrane three times by incubating in 10 mL of TBS-T with rocking for 10 min each time.

9. Prepare secondary antibody dilutions in TBS-T-BSA1 dilution buffer (1/2500 for HRP-anti-mouse IgG antibodies and 1/5000 for HRP anti-rabbit IgG antibodies). Typically, 8–10 mL of antibody solution is enough to cover surface of a membrane in "mini-gel" format.
10. Incubate membrane in secondary antibodies solution for 1 h at room temperature with rocking.
11. Wash three times the membrane by incubating in 10 mL of TBS-T for 10 min with rocking each time.

3.7 Enhanced Chemiluminescence Imaging

1. Prepare ECL solution by mixing the two solutions at a 1:1 ratio (for a typically sized "mini-gel" format membrane ~3–5 mL of ECL solution should be enough to cover area of a mini-gel).
2. Blot-out excess moisture from the membrane with Kimwipe tissue by placing one edge of the membrane on the tissue. Do not let the surface of a membrane contact directly tissues or other absorbing surfaces.
3. Place the membrane on a flat surface with the side that has contacted the gel facing up.
4. Add a few drops of ECL solution mixture to cover the entire surface of the membrane.
5. Incubate for 1 min at room temperature.
6. Place membrane in ECL imager (*see* **Note 16**).
7. Turn imager device on and proceed to imaging in normal light mode (for ladder) and chemiluminescence mode (ECL).
8. Transfer image files on a computer and proceed with western blot image analyses. Representative results of this method are shown in Fig. 3 and a short summary of results are found in **Note 17**.

4 Notes

1. This protocol allows characterization of spike protein proteolytic processing by transient co-transfection of MERS-CoV S-encoding and human matriptase-encoding plasmids. However, there are other methods to analyze the envelope glycoprotein cleavage step. For example, using murine hepatitis virus (MHV) coronavirus infection in susceptible cells, Wicht and colleagues have elegantly engineered a biotinylation-based intracellular labeling assay to selectively label spike proteins that have undergone fusion to analyze their cleavage status [24]. Note that in the case of MERS-CoV infection, the latter experiment would require access to a BSL-3 facility. Another, more biochemical approach is to incubate viral particles such as

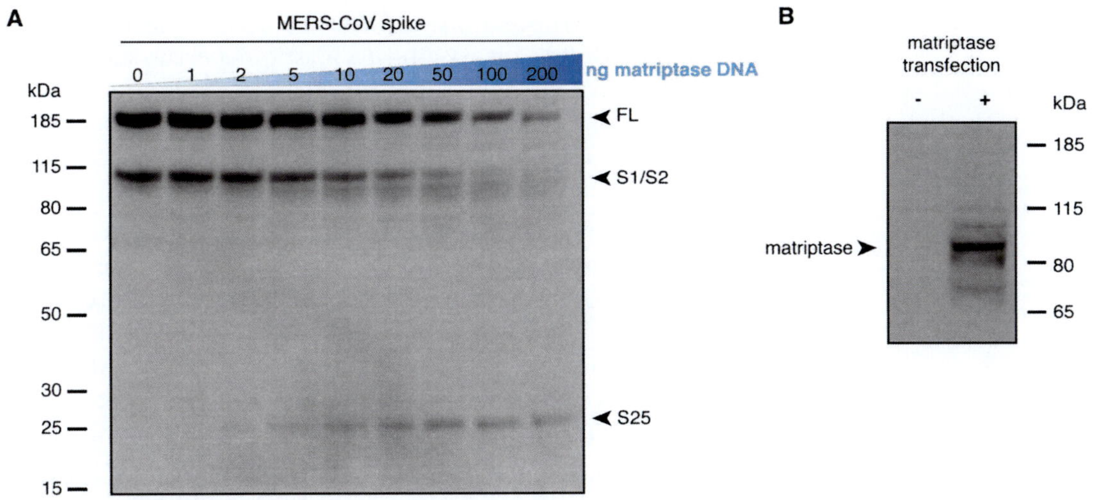

Fig. 3 Proteolytic processing of MERS-CoV spike protein by matriptase expression. (**a**) Spike protein western blot analysis. HEK-293 T/17 cells are co-transfected with MERS-CoV spike-encoding and human matriptase-encoding plasmids. The amount of MERS-CoV spike-encoding plasmid is kept constant while the quantity of human matriptase encoding plasmid is gradually increased from 0 to 200 ng as shown on the blue scale triangle. 24 h post-transfection cells are lysed, lysates processed and analyzed by western blot using rabbit polyclonal antibodies raised against MERS-CoV spike protein ectodomain (see note 17 for a summary of the results obtained). FL: full-length spike protein; S1/S2: S1/S2-cleaved fragment. (**b**) Matriptase expression in transfected cells. To check for protease expression in HEK-293 T/17 cells after transfection of matriptase-encoding plasmid (+, 500 ng) or pcDNA3.1 empty vector control (−, 500 ng), transfected cells were lysed and analyzed by western blot using mouse monoclonal antibodies specific for detecting human matriptase. A strong band was detected for the condition where the human matriptase-encoding plasmid was used for transfection. This band migrated at approximately 90 kDa, which corresponds to the expected apparent molecular weight of human matriptase

coronavirus spike-pseudotyped particles with a purified recombinant protease and perform a western blot assay [17, 25].

2. In our hands, the HEK-293 T/17 cells described in this protocol are found to be only semi-adherent and are prone to easily detach from cell culture surfaces. If this issue is encountered, we recommend to handle the cells with care, being mindful to use pre-warmed (37 °C) DPBS and media as much as possible. We have found that pretreating cell culture plastic surfaces with poly-D-lysine helps the cells adhere better to surfaces. During cell detachment we recommend avoiding incubating cells with trypsin for more than 5 min as this paradoxically often leads to cells aggregating.
3. To avoid cells clumping in the middle of wells and being sparsely distributed near the edges, we recommend to refrain from circular motion or tilting the plate.
4. In this protocol we use Lipofectamine 2000 as transfection reagent. We found that it works well for the cell type being used (HEK-293 T/17). Other transfection reagents and/or

methods (electroporation) can also be used, if needed. Lipofectamine 2000 tends to bind to plastic materials such as the walls of microcentrifuge tubes. As such, it is recommended to always place Lipofectamine 2000-containing solutions in tubes that already contain a diluent rather than in direct contact with the walls of microcentrifuge tubes.

5. Transfection reagents such as the one used here (Lipofectamine 2000) increase cell membrane permeability. Because of this effect, antibiotics that are typically added in cell culture media (penicillin and streptomycin) tend to increase in cytotoxicity. It is thus recommended to use cell medium devoid of antibiotics after cells receive transfection reagents such as Lipofectamine 2000.

6. LDS stored at 4 °C partially precipitates. To better pipet correct volumes, we recommend warming LDS at 37 °C to ensure working with a correctly solubilized solution. Also, we have found for other coronavirus spike proteins that reducing conditions (DTT) and/or application of heat (95 °C for 5 min) can negatively impact antibody detection at later steps of the western blot assay. We suggest to test different conditions (with or without DTT and/or with or without heat) to check which ones work best for a given coronavirus spike protein.

7. At this point, protein samples can be directly analyzed by western blot assay or stored at −20 °C until used. Once frozen and kept at −20 °C, the samples are stable for several months. The protein samples are relatively tolerant to a few freeze-thaw cycles but we would suggest to refrain from multiple freeze-thaw cycles (5 or more). A workaround to avoid this problem is to aliquot samples into smaller volumes prior to the initial freezing.

8. The polyacrylamide gel used in this protocol is a pre-cast 4–12% gradient gel. We have found that it allows to obtain good separation of full-length coronavirus spike and fragments generated by proteolytic activity. Other kinds of gels can be used as well, including gels made in the lab, which are relatively easy to prepare.

9. Most of the gel migration is performed at 200 V with constant voltage. However, we have sometimes found that applying the full voltage of 200 V from the start of migration can result in irregularities in the migration pattern of protein bands. To avoid this, we have found that starting at a lower voltage (e.g., 100 V) and incrementally increasing (e.g., 20 V increments) to 200 V over a 10–15 min timeframe allows to obtain more consistent protein migration.

10. Here, ethanol can be used in place of methanol.

11. In this chapter a PVDF membrane is used. Alternatively, a nitrocellulose membrane can also be used as we have found that such membrane type can also allow transfer of coronavirus spike protein bands.
12. Gels and membranes dry easily. As much as possible, avoid leaving them to dry in air and handle them within containers with buffer solution.
13. Due to their size and heavy glycosylation status, full-length MERS-CoV spike proteins (and most other coronavirus spike proteins) are relatively large with monomers typically migrating with apparent molecular weight of around 180 kDa or more and require relatively long transfer times when performing "wet" transfers as presented here. We found that 3 h allows to properly transfer the protein from the gel to the PVDF membrane. This long transfer time typically leads to heating of the transfer tank and transfer solution. To avoid overheating and potential protein degradation, in addition to placing an ice pack in the transfer tank and using prechilled transfer buffer, we recommend encasing the tank within an ice bucket filled with ice. Another option is to perform the transfer in a refrigerated setting such as within a cold room.
14. Here, BSA is used for preparing blocking solution (in TBS-T) as well as the diluent for antibody incubations. We have found that BSA at concentrations of 1–2% generally gives clear western blot images with low background signal. An alternative to BSA is to use powdered milk. The concentrations and conditions for incubations with TBS-T supplemented with powdered milk should be tested beforehand to check which one gives the best results in terms of western blot background.
15. The dilutions and incubation times for the antibodies used in this protocol were chosen after testing different conditions in preliminary experiments. If other antibodies are used, we suggest to also perform such tests to determine which antibody dilutions and incubation times give the best results.
16. The imaging method presented in this protocol uses a chemiluminescence imager. However, it's also possible to perform this using X-ray film and developer.
17. The assay presented here confirms that in the absence of matriptase expression (Fig. 3b, lane -), the MERS-CoV spike protein is cleaved at the S1/S2 site by an intracellular protease (Fig. 3a, lane 0), most likely of the pro-protein convertase family (e.g., furin), as shown previously [17]. The analysis also reveals that upon expression of matriptase, there is dose-dependent increase in the detection of a MERS-CoV spike protein proteolytic fragment (S25) migrating at approximately 25 kDa (Fig. 3a, lanes 1–200). In addition, the band signals

corresponding to full-length (FL) and S1/S2-cleaved fragment (S1/S2) appear to decrease upon increased matriptase expression, a result which could indicate that high levels of matriptase expression (e.g., in the 100–200 ng range) are associated with nonspecific degradation of MERS-CoV spike protein. This effect could also be due to the matriptase and MERS-CoV spike co-transfection conditions used here, and it would be interesting to compare these results with the proteolytic processing of matriptase during viral entry. In addition, matriptase is expressed in this system without its natural inhibitor, HAI-1, which potently regulates the enzyme's activity. It is also noteworthy to point out that even in conditions with very low amounts of transfected matriptase-encoding plasmid (e.g., 2 ng) the S25 band is detected, suggesting that matriptase may be highly processive for the MERS-CoV spike substrate, in the absence of inhibitor. The composition of S25, the associated matriptase cleavage site and the functional consequence of such cleavage await further elucidation.

Acknowledgments

We wish to thank all members of the Whittaker lab for helpful comments. We gratefully acknowledge Marco Straus for providing the plasmid which encodes the sequence for human matriptase. Research funding for this work was provided by the NIH grants R21 AI111085 and R01 AI135270.

References

1. Fehr AR, Perlman S (2015) Coronaviruses: an overview of their replication and pathogenesis. Methods Mol Biol 1282:1–23
2. Bosch BJ, Rottier P (2008) Nidovirus entry into cells. In: Perlman S, Gallagher T, Snijder EJ (eds) Nidoviruses. ASM Press, Washington, DC, pp 157–178
3. Bosch BJ, van der Zee R, de Haan CAM, Rottier PJM (2003) The coronavirus spike protein is a class I virus fusion protein: structural and functional characterization of the fusion Core complex. J Virol 77(16):8801–8811
4. White JM, Delos SE, Brecher M, Schornberg K (2008) Structures and mechanisms of viral membrane fusion proteins: multiple variations on a common theme. Crit Rev Biochem Mol Biol 43(3):189–219
5. Millet JK, Whittaker GR (2015) Host cell proteases: critical determinants of coronavirus tropism and pathogenesis. Virus Res 202:120–134
6. Belouzard S, Chu VC, Whittaker GR (2009) Activation of the SARS coronavirus spike protein via sequential proteolytic cleavage at two distinct sites. Proc Natl Acad Sci 106 (14):5871–5876
7. White JM, Whittaker GR (2016) Fusion of enveloped viruses in endosomes. Traffic 17 (6):593–614
8. Park JE et al (2016) Proteolytic processing of Middle East respiratory syndrome coronavirus spikes expands virus tropism. Proc Natl Acad Sci U S A 113(43):12262–12267
9. Lazarowitz SG, Choppin PW (1975) Enhancement of the infectivity of influenza a and B viruses by proteolytic cleavage of the hemagglutinin polypeptide. Virology 68(2):440–454
10. Lazarowitz SG, Compans RW, Choppin PW (1973) Proteolytic cleavage of the hemagglutinin polypeptide of influenza virus. Function of the uncleaved polypeptide HA. Virology 52 (1):199–212

11. Klenk HD, Matrosovich M, Stech J (2008) Avian influenza: molecular Mechansims of pathogenesis. In: Mettenleiter T, Sobrino F (eds) Animal Viruses: Molecular Biology. Caister Academic Press, Norfolk, UK, pp 253–303
12. Gierer S et al (2013) The spike-protein of the emerging betacoronavirus EMC uses a novel coronavirus receptor for entry, can be activated by TMPRSS2 and is targeted by neutralizing antibodies. J Virol 87(10):5502–5511
13. Qian Z, Dominguez SR, Holmes KV (2013) Role of the spike glycoprotein of human Middle East respiratory syndrome coronavirus (MERS-CoV) in virus entry and syncytia formation. PLoS One 8(10):e76469
14. Shirato K, Kawase M, Matsuyama S (2013) Middle East respiratory syndrome coronavirus (MERS-CoV) infection mediated by the Transmembrane serine protease TMPRSS2. J Virol 87(23):12552–12561
15. Kawase M, Shirato K, van der Hoek L, Taguchi F, Matsuyama S (2012) Simultaneous treatment of human bronchial epithelial cells with serine and cysteine protease inhibitors prevents severe acute respiratory syndrome coronavirus entry. J Virol 86(12):6537–6545
16. Burkard C et al (2014) Coronavirus cell entry occurs through the Endo−/Lysosomal pathway in a proteolysis-dependent manner. PLoS Pathog 10(11):e1004502
17. Millet JK, Whittaker GR (2014) Host cell entry of Middle East respiratory syndrome coronavirus after two-step, furin-mediated activation of the spike protein. Proc Natl Acad Sci 111 (42):15214–15219
18. Matsuyama S et al (2018) Middle East respiratory syndrome coronavirus spike protein is not activated directly by cellular Furin during viral entry into target cells. J Virol 92(19):e00683-18
19. Kleine-Weber H, Elzayat MT, Hoffmann M, Pöhlmann S (2018) Functional analysis of potential cleavage sites in the MERS-coronavirus spike protein. Sci Rep 8(1):16597
20. Oberst MD et al (2003) Characterization of matriptase expression in normal human tissues. J Histochem Cytochem 51(8):1017–1025
21. Shimomura T et al (1997) Hepatocyte growth factor activator inhibitor, a novel Kunitz-type serine protease inhibitor. J Biol Chem 272 (10):6370–6376
22. Beaulieu A et al (2013) Matriptase proteolytically activates influenza virus and promotes multicycle replication in the human airway epithelium. J Virol 87(8):4237–4251
23. Hamilton BS, Gludish DW, Whittaker GR (2012) Cleavage activation of the human-adapted influenza virus subtypes by Matriptase reveals both subtype and strain specificities. J Virol 86(19):10579–10586
24. Wicht O et al (2014) Identification and characterization of a Proteolytically primed form of the murine coronavirus spike proteins after fusion with the target cell. J Virol 88 (9):4943–4952
25. Millet JK et al (2016) A camel-derived MERS-CoV with a variant spike protein cleavage site and distinct fusion activation properties. Emerg Microbes Infect 5(12):e126

Chapter 4

Crystallization and Structural Determination of the Receptor-Binding Domain of MERS-CoV Spike Glycoprotein

Haixia Zhou, Shuyuan Zhang, and Xinquan Wang

Abstract

Three-dimensional structures of the receptor-binding domain (RBD) of MERS-CoV spike glycoprotein bound to cellular receptor and monoclonal antibodies (mAbs) have been determined by X-ray crystallography, providing structural information about receptor recognition and neutralizing mechanisms of mAbs at the atomic level. In this chapter, we describe the purification, crystallization, and structure determination of the MERS-CoV RBD.

Key words X-ray crystallography, MERS-CoV, Spike, RBD, Crystallization, Structure determination

1 Introduction

The first three-dimensional structure of the MERS-CoV spike glycoprotein receptor-binding domain (RBD), providing the molecular basis of viral attachment to host cells, was determined in the complex with it cellular receptor dipeptidyl peptidase 4 (DPP4, also called CD26) by X-ray crystallography [1]. Because of the significance in receptor recognition and specific pathogenesis, RBD became a hot spot in the study of MERS-CoV. A number of structures of RBD bound by monoclonal antibodies (mAbs) have also been determined and deposited in the Protein Data Bank (PDB, http://www.rcsb.org/pdb/) [2–8]. Our group determined the RBD structures in complex with DPP4 and the mAbs MERS-27, MERS-4 and MERS-GD27, respectively [2–4, 9]. All the three-dimensional structures of MERS-CoV RBD have been determined by X-ray crystallography, which is a powerful method for determining molecular structures at atomic resolution. Briefly, the ordered and repeated atoms in a single protein crystal can diffract the incident X-ray beam into many specific directions. The angles and intensities of these diffracted X-rays can be collected and

Rahul Vijay (ed.), *MERS Coronavirus: Methods and Protocols*, Methods in Molecular Biology, vol. 2099,
https://doi.org/10.1007/978-1-0716-0211-9_4,

measured in an X-ray diffraction experiment. After obtaining the phases of these diffracted X-rays by heavy-atom derivative, anomalous scattering or molecular replacement methods, a protein crystallographer then calculates the density of electrons with the protein crystal and builds a structural model based on the density map. For details on the principles and methodology of protein crystallography, please refer to the range of other excellent textbooks.

In this chapter, an overview of the standard method of protein crystallography is briefly introduced, focusing on crystallization and structural determination of MERS-CoV RBD using the molecular replacement method.

2 Materials

Prepare all solutions using ultrapure water (prepared by purifying deionized water, to attain a resistivity of 18 MΩ cm at 25 °C) and analytical grade reagents. When dealing with waste, we strictly follow all waste disposal regulations.

2.1 Expression

1. pFastBac vector containing the MERS-CoV RBD gene.
2. DH10Bac competent cells.
3. LB liquid medium: 10 g tryptone, 5 g yeast extract, 10 g NaCl, and 1 L ultrapure water; sterilize by high-pressure steam.
4. Liquid LB selection medium: LB liquid medium, 50 μg/mL kanamycin, 7 μg/mL gentamicin, and 10 μg/mL tetracycline.
5. Bacmid selection LB agar plate:10 g tryptone, 5 g yeast extract, 10 g agar powder, 10 g NaCl, and 1 L ultrapure water. Sterilization at high-pressure steam. 50 μg/mL kanamycin, 7 μg/mL gentamicin, 10 μg/mL tetracycline, 100 μg/mL x-gal, 40 μg/mL IPTG Mix and pout into sterile plates (*see* **Note 1**).
6. TIANprep Mini Plasmid Kit (TIANGEN); Buffers P1, P2, and P3 (*see* **Note 2**).
7. Isopropanol.
8. 70% Ethanol.
9. Sf9 cell line.
10. Sf-900 II Serum-Free Medium.
11. Cellfectin II Reagent (store at 4 °C).
12. Cell Culture Dish.

2.2 Purification

1. 10× HBS buffer: 100 mM HEPES, 1500 mM NaCl, water (adjust pH to 7.2 with NaOH) (*see* **Note 3**).

2. 1× HBS buffer: 100 mL 100× HBS buffer, 900 mL water (*see* **Note 3**).
3. Wash buffer: 10 mM HEPES, pH 7.2, 150 mM NaCl, 20 mM imidazole (*see* **Note 4**).
4. Elution buffer: 10 mM HEPES, pH 7.2, 150 mM NaCl, 500 mM imidazole.
5. Coomassie brilliant blue G-250 (Solarbio; *see* **Note 5**).
6. Ni-NTA Resin (GE Healthcare).
7. Endoglycosidases F1 and F3; store at 4 °C (*see* **Note 6**).
8. Crossflow ultrafiltration system.
9. Vivaspin Turbo ultrafiltration spin column: membrane 30,000 MWCO PES.

2.3 Crystallization

1. Crystallization kits: Crystal Screen, Index, PEG/Ion, PEGRx, SaltRx, Natrix from Hampton Research; Structure Screen, Proplex from Molecular Dimension; Wizard I, Wizard II from Emerald Biosystems; JCSG+Suit from QIANGEN.
2. SWISSCI 3 Lens crystallization plate.
3. Mosquito crystallization setups.
4. 8-Well 5 μL micro-reservoir strip.

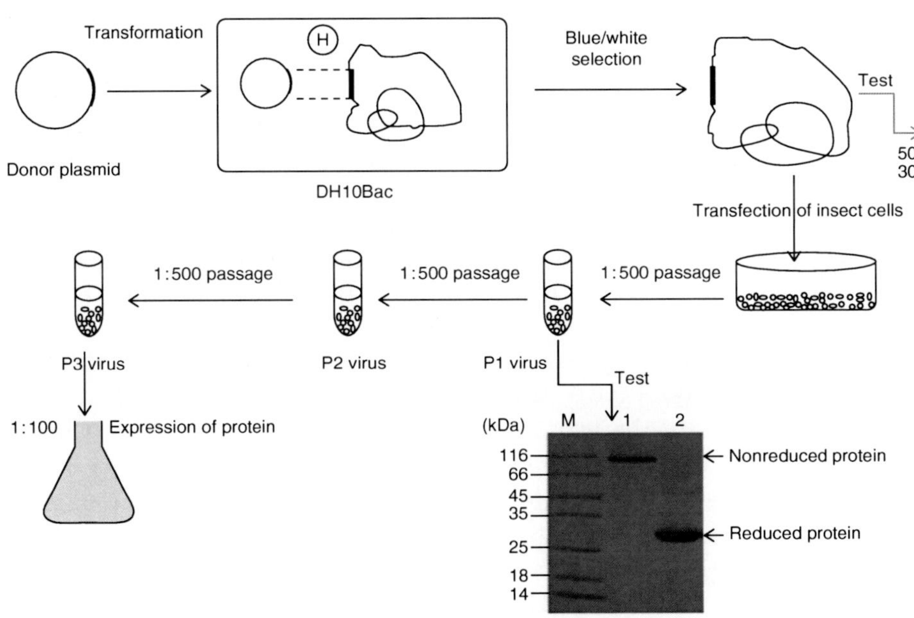

Fig. 1 Schematic diagram of bac-to-bac expression system

3 Methods

MER-CoV RBD can be expressed using the Bac-to-Bac baculovirus expression system (Fig. 1), collected and captured using NTA Sepharose (GE Healthcare) and then further purified by gel filtration chromatography using a Superdex 200 High Performance column (GE Healthcare). Crystallization trials are set up using the hanging-drop or sitting-drop vapor diffusion method in conjunction with the sparse-matrix crystal screening kits. The structure of MERS-CoV RBD in complex with MERS-4scFv was determined using the molecular replacement method.

3.1 Construction of Recombinant Baculovirus

1. Transforming the DH10Bac cells with the plasmid containing the MERS-CoV RBD gene (*see* **Note 7**): add the recombinant plasmid to 50 μL of the DH10Bac competent cells. Incubate on ice for 30 min. Heat-shock the cells for 45 s at 42 °C without shaking. Immediately transfer the tubes onto ice and chill for 2 min. Add 500 μL of room temperature LB medium. Shake the tubes at 37 °C in incubator shaker for 4 h and then plate 20 μL of the cell suspension onto a bacmid selection LB agar plate. Incubate plates for 24–48 h at 37 °C, dark.
2. Isolation of recombinant bacmid DNA: Pick a white colony from the bacmid selection LB agar plate, and inoculate 3 mL of liquid LB selection medium with the picked colony. Harvest the cells by centrifuging at 3000 × *g* for 10 min, and then remove the entire medium. Add 250 μL of buffer P1 containing RNase A to the pellet, and resuspend the cells until the suspension is homogeneous. Transfer the cell suspension to a centrifuge tube. Add 250 μL buffer P2 and mix gently by inverting the capped tube several times. Incubate at room temperature for 5 min. Add 350 μL Buffer P3 and mix immediately by inverting the capped tube until the mixture is homogeneous. Centrifuge the mixture at 15,000 × *g* at room temperature for 10 min. Transfer 800 μL of the supernatant into a 2 mL tube with 500 μL pre-chilled isopropanol at −20 °C for 20 min. Centrifuge the mixture at 15,000 × *g* at 4 °C for 20 min. Carefully discard the supernatant. Resuspend the DNA pellet in 1 mL of 70% ethanol. Centrifuge at 15,000 × *g* at 4 °C for 5 min. Carefully discard the supernatant. Air-dry the pellet for 30 min. Resuspend the DNA pellet in 50 μL of sterile ultrapure water (*see* **Note 8**).
3. Confirm the bacmid quality by separating the DNA on a 0.5% agarose gel. The bacmid is in the position nearest the gel hole. Verify the presence of the MERS-CoV RBD gene in the recombinant bacmid by PCR, using M13 forward and reverse primers

(M13 F: CCCAGTCACGACGTTGTAAAACG; M13 R: AGCGGATAACAATTTCACACAGG; optional).

4. Transfection of Sf9 cells: verify that the Sf9 cells are in the log phase (1.5–2.5 × 10^6 cells/mL) with greater than 95% viability. Add 2 × 10^6 cells into a 10 cm dish with 2 mL of culture medium (Sf-900 II Serum Free). Allow the cells to attach themselves for 30–60 min at room temperature in the hood. For the transfection sample, dilute 10 μL Cellfectin II Reagent in 250 μL of culture medium (Sf-900 II Serum Free). Mix by inverting 5–10 times (do not vortex), and incubate for 5 min at room temperature. Add 1 μg of bacmid DNA. Mix by inverting 5–10 times (do not vortex), and incubate for 20 min at room temperature. Add the entire DNA-lipid mixture dropwise onto the cells. Incubate the cells at 27 °C for 5 h and then remove the supernatant and add 7 mL of fresh culture medium (Sf-900 II Serum Free).
5. After 7–9 days or until visible signs of virus infection (*see* **Note 9**), transfer the medium containing the virus to sterile 15-mL tubes. Centrifuge the tube at 600 × g for 5 min to remove cells and large debris. Transfer this clarified supernatant to a fresh 15-mL tube.
6. Store this P0 virus stock at 4 °C, protected from light (*see* **Note 10**).
7. Amplifying the baculovirus stock: On the day of infection, prepare an Sf9 cell suspension by seeding at 2 × 10^6 cells/mL in 50 mL of culture medium. Add 100 μL of P0 virus stock to the flask. Incubate the suspended cells for 72–96 h at 27 °C in incubator shaker. Transfer the medium containing the virions from the flask to a sterile 50-mL tube, and centrifuge the tube at 600 × *g* for 5 min to clarify the baculovirus stock. Transfer the supernatant to a fresh collection tube. This is the P1 virus stock.
8. Store this P1 virus stock at 4 °C, protected from light. The P2 and P3 virus can be obtained by repeating **step 7**.

3.2 Protein Expression and Purification

1. The high-titer P2 baculovirus stock was used to infect 2 × 10^6 Sf9 cells per milliliter. The baculovirus virions are generally added according to the virus to cell volume ratio of 1:100 and then the Sf9 cells are cultured at 27 °C in incubator shaker for 48–60 h.
2. After transfection for 48–60 h, collect the cell-culture supernatant containing the MERS-CoV RBD by centrifugation at 3000 × *g* for 15 min to remove cells, and pass the supernatant through a 0.45 μm filter, to avoid blocking the filter of the crossflow ultrafiltration system.

3. For subsequent purification, concentrate large volumes of supernatant to a smaller volume. We used the Sartorius cross-flow ultrafiltration system for concentration. When the volume is concentrated to about 50 mL, add 200 mL portions of HBS buffer (1 L in total) to the big beaker to exchange the buffer. When the total volume remaining is reduced to 50 mL, add 100 mL HBS buffer to collect all the liquid in the system in a beaker. Then dispense into high-speed centrifuge tubes, centrifuge at 3000 × *g* for 1 h at 4 °C.
4. The supernatant after centrifugation is loaded onto the nickel-NTA beads equilibrated with 30 mL of HBS buffer. MERS-CoV RBD with a His tag could be captured by nickel-NTA beads. Repeat loading the sample once more.
5. Add wash buffer to the beads to remove the nonspecifically bound proteins until the flow-through is not able to discolor the Coomassie Brilliant Blue G250 solution.
6. After adding elution buffer, the target protein will dissociate from the beads; collect it in a 10 kDa Millipore concentrating tube. Similarly, detect protein with Coomassie Brilliant Blue G250. The concentrating tube containing the protein sample is centrifuged at 3000 × *g* to concentrate the sample to less than

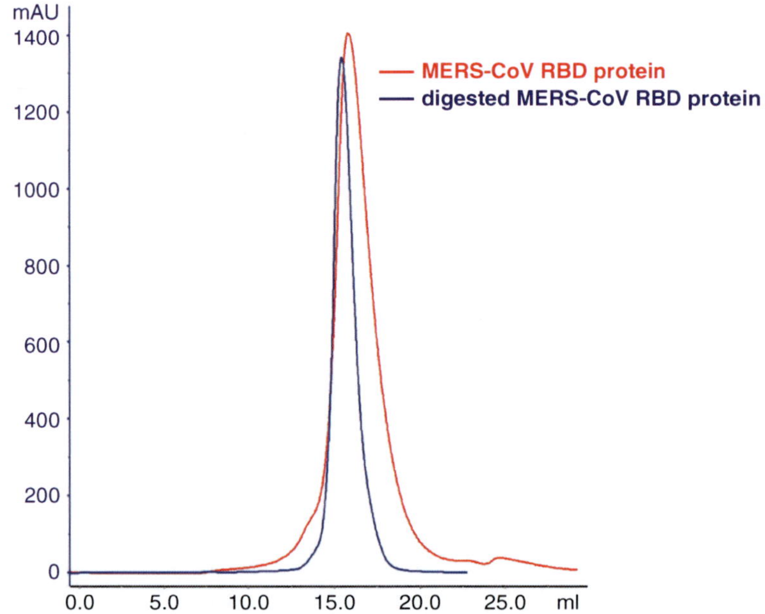

Fig. 2 Gel filtration profile of MERS-CoV RBD protein and digested by endoglycosidase F1 and F3 MERS-CoV RBD protein, confirmed by SDS-PAGE. The first two lanes are MERS-CoV RBD protein and the last two lanes are digested RBD protein

1 mL. Transfer the sample to a 1.5 mL EP tube and centrifuge at 3000 × *g* for 10 min.

7. MERS-CoVRBD is further purified by gel filtration chromatography. The sample is loaded onto the Superdex 200 column pre-equilibrated with HBS buffer. Fractions containing RBD are collected and the protein's purity is confirmed by SDS-PAGE (Fig. 2).
8. Dilute the protein to 1 mg/mL, and digest with endoglycosidase F1 and F3 (F1/F3: RBD at the ratio of 1:100) at 18 °C overnight. The digested protein is concentrated and purified by gel filtration chromatography same as above (optional).
9. After preparing the MERS-CoV RBD protein and MERS-4scFv protein (*see* **Note 11**) detect the absorption of the protein sample at 280 nm (A280). According to the molecular weight and extinction coefficient, the molar concentration can be calculated. The two proteins were mixed at molar ratio of 1:1, incubated on ice for 1 h, and purified using a Superdex 200 column. Collect the fractions containing the complex and confirm the protein purity by SDS-PAGE.

3.3 Crystallization

1. Collect the purified MERS-CoV RBD protein and use a 2 mL Sartorius centrifugal concentrator to concentrate to 10–15 mg/mL. After mixing and aspirating, centrifuge at 10,000 × *g* for 10 min at 4 °C.
2. Use commercial reservoir solution kits for crystallization. Every kit has at least 96 different conditions, including salt, buffer, precipitant, and pH. Dispense 30 μL reservoir solution onto the 96-well SWISSCI3 Lens crystallization plate.
3. Use TTP LabTech's mosquito crystallization setup for automated crystallography. Absorb 3 μL protein on the 8-well 5 μL micro-reservoir strip (Fig. 3a). Then the needles aspirate the protein from the strip onto the SWISSCI plate with 200 nL of protein per well (Fig. 3b), using the sitting-drop vapor diffusion method by mixing 200 nL reservoir and 200 nL reservoir (Fig. 3c,d).
4. Seal the plate with tape and gently place it in an 18 °C room.
5. Check the sample drops under a microscope at 20–100× magnifications after 3 and 7 days (and if necessary, after 1 and 2 weeks, and 1, 3, and 6 months).
6. A week later, we found crystal growth in the PEG/Ion, PEGRX and JCSG+ kits. Specific conditions were as follows (Fig. 4a,b). PEG/ION:0.2 M Potassium sodium tartrate tetrahydrate, 20% w/v polyethylene glycol 3350.PEGRx:0.1 M BICINE pH 8.5, 20% w/v polyethylene glycol 10,000. JCSG +: 0.2 M Potassium nitrate, 20%(w/v) PEG 3350.

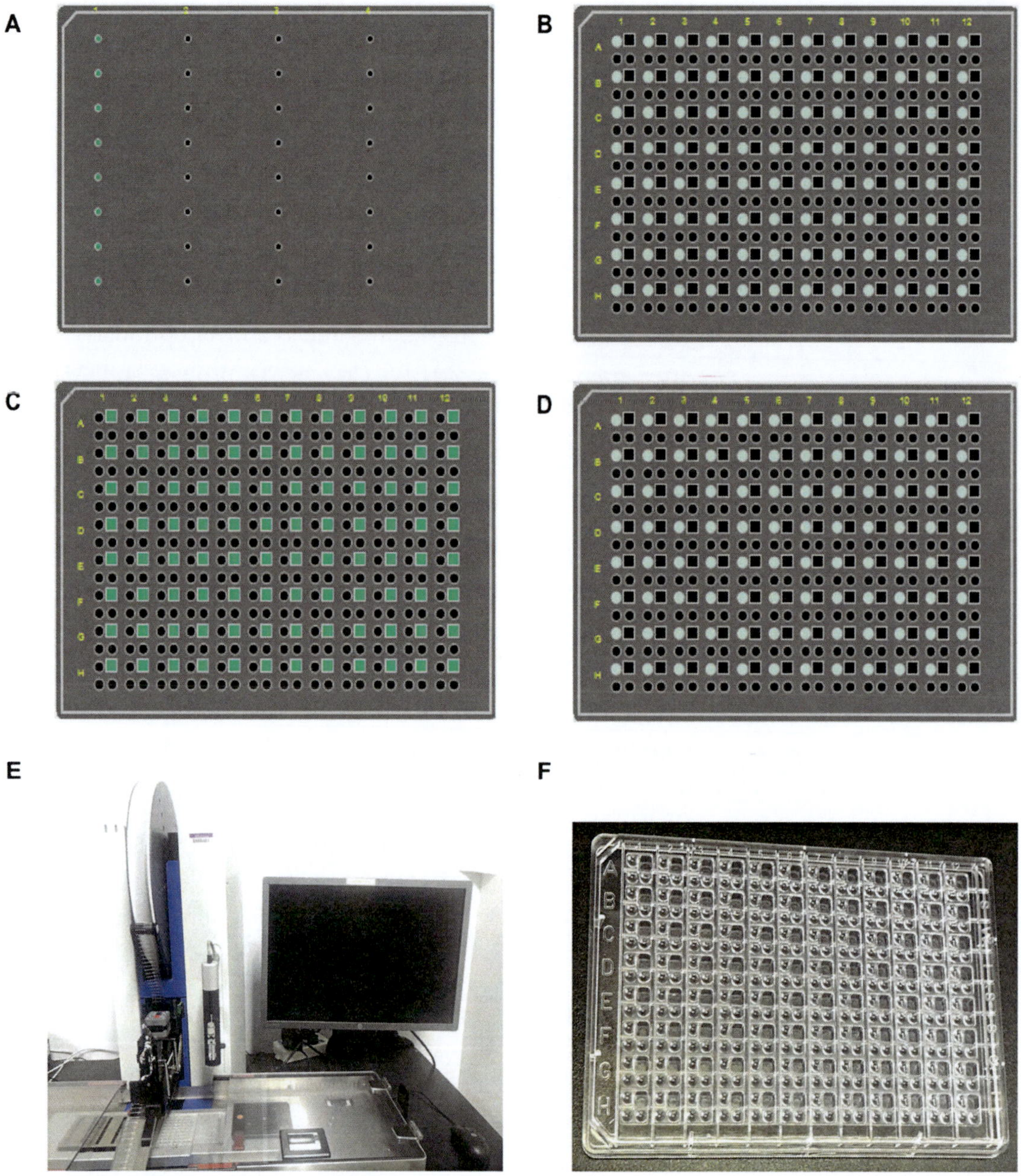

Fig. 3 Crystallization by mosquito. Panel **a–d** shows the operation diagram of the TTP LabTech's setups: blue dots represent proteins, and square lattices represent reservoir. Panel **e** shows the TTP LabTech's Mosquito machine. Panel **f** shows the SWISSCI plate

3.4 Structural Determination

1. All crystals should be flash-frozen in liquid nitrogen after being incubated in the reservoir solution containing 20% (v/v) glycerol.

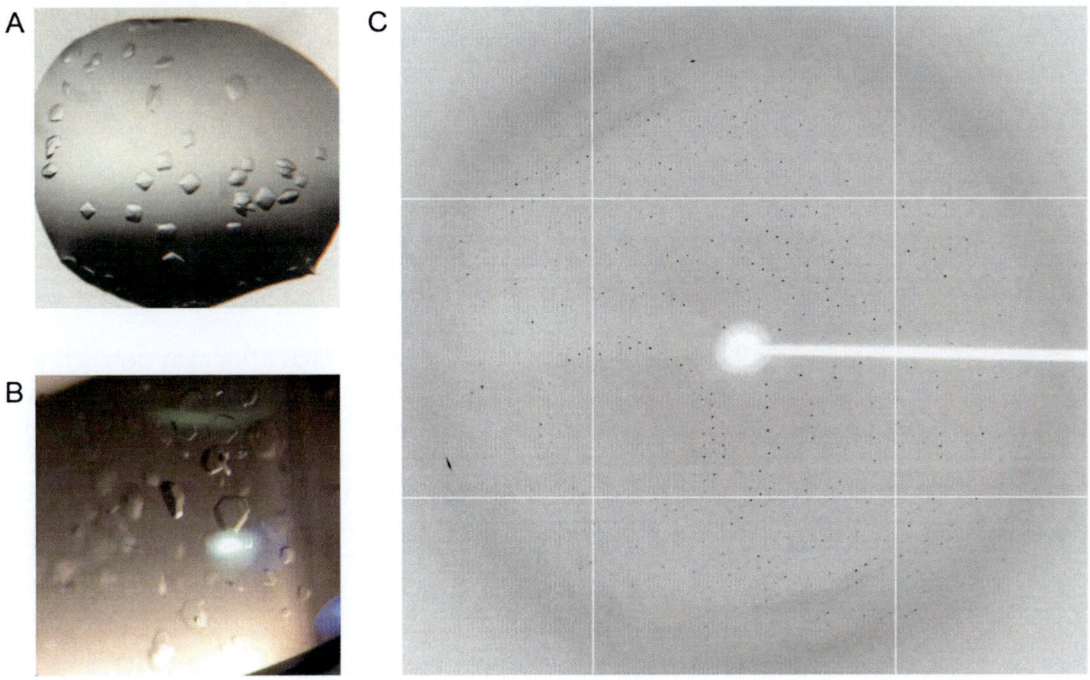

Fig. 4 Crystals of MERS-CoV RBD and MERS-4scFv complex (**a**, **b**) and the X-ray diffraction pattern of the crystal complex (**c**)

2. The diffraction images should be collected on the BL17U beamline (Fig. 4c). Rotate the mounted crystal and the X-ray diffraction patterns should be recorded at 1° per image, and collected for 360°.

3. The diffraction images in a dataset should be processed with HKL2000 [10] including auto-index, refinement, integration, and scaling steps. After data processing, the crystal unit cell parameters, crystal space group, Miller indexes of reflections, intensities, and error estimates of reflections should be determined and stored in a ∗.sca file, which provides the dataset applicable to structure determination.

4. Using CCP4 suite solve the structure as follows: Export the ∗.sca file to a ∗.mtz file using the program SCALPACK2MTZ. Use the ∗.mtz file to calculate the solvent via MATTHEWS_COEF. Run with PHASER MR (*see* **Note 12**) with the MERS-CoV RBD structure (PDB ID: 4l72) and the structures of the variable domain of the heavy and light chains available in the PDB with the highest sequence identities as search models (*see* **Note 12**) [11]. When the phases are determined, the electron density map can be calculated, from which the molecular model can be constructed.

5. Subsequent model building and refinement were performed using COOT and PHENIX, respectively (*see* **Note 13**) [12, 13].
6. Many validation programs are used to check the structure, until the investigator is satisfied, and then the structure can be deposited in the PDB.

4 Notes

1. Weigh 10 g tryptone, 5 g yeast extract, 10 g agar powder, and 10 g NaCl, and add ultrapure water to 1 L. After sterilization by high-pressure steam, wait until the temperature of the medium drops to about 60 °C, add the required antibiotics, inducers etc. (50 μg/mL kanamycin,7 μg/mL gentamicin, 10 μg/mL tetracycline, 100 μg/mL x-gal, 40 μg/mL IPTG). Mix evenly and then pour into sterile plates. When the culture medium has cooled and solidified, store the bacmid selection LB agar plate at 4 °C.
2. TIANprep Mini Plasmid Kit includes Buffers P1, P2, and P3. We only use the Buffers P1, P2, and P3 for the preceding steps of isolating recombinant bacmid DNA.
3. Method for preparing HBS buffer: Prepare 10× HBS buffer (100 mM HEPES, 1500 mM NaCl). Weigh 23.8 g HEPES (Sigma), 87.66 g NaCl (Amresco) dissolve with water to 1 L, adjust pH to 7.2 with NaOH. Then, dilute 100 mL of 10× HBS buffer with 900 mL of water.
4. Method of preparing wash buffer: First prepare 5 M imidazole. Weigh 340 g imidazole (Sigma-Aldrich, USA), add water to 1 L and use HCl to adjust pH to 8.0. Add 100 mL 10× HBS buffer, 30 mL 5 M imidazole with water to 1 L.
5. Method of preparing Coomassie brilliant blue G-250: Weigh 1 g G-250 (Solarbio), add 100 mL 85% phosphate, 100 mL anhydrous ethanol and water to 1 L. The G-250 needs to be stored away from light.
6. Endoglycosidase F1 and F3 are expressed and purified from *E. coli* by our laboratory. The endoglycosidase was added into the reaction system according to the mass ratio of 1:100.
7. The coding sequence of the MERS-CoV RBD (EMC strain, spike residues 367–588) was ligated into the pFastBac-Dual vector (Invitrogen) with a N-terminal gp67 signal peptide to enable the protein secreting outside the cell and a C-terminal His-tag to facilitate further purification processes.
8. Allow the pellet to dissolve for at least 10 min at room temperature. To avoid shearing the DNA, pipet only 1–2 times to

resuspend. Store the bacmid at 4 °C and use it as soon as possible, usually within 1 week. Aliquot the bacmid DNA into separate tubes and store at −20 °C (not in a frost-free fridge). Avoid multiple freeze/thaw cycles as this decreases the transfection efficiency.

9. Characteristics of infected cells: A 25–50% increase in cell diameter can be seen and the size of cell nuclei increases at the early stage. Cells release from the plate and appear lysed, showing signs of clearing in the monolayer.
10. P0 virus can be stored for years, adding 2% (v/v) FBS at 4 °C, protected from light.
11. The expression of MERS-4scFv protein was conducted in 293F cells transiently transfected with plasmid DNA. After 72 h, the supernatant was collected and concentrated. The purified MERS-4scFv protein was obtained by Ni-NTA affinity chromatography and Superdex 200 size-exclusion chromatography. The purification method is the same as that of RBD protein.
12. If the crystal structure of the same protein or a similar protein has been solved, the molecular replacement method can be applied. After obtaining the solutions of the rotation and translation functions, initial phases can be calculated from the reference model, after which the electron density can be calculated.
13. The accuracy of the constructed model is confirmed by the crystallographic R-factor and R-free, which indicate the discrepancy between the calculated and observed amplitudes. The stereo-chemical parameters of the model can also be checked using programs such as MOLPROBITY, PROCHECK, or RAMPAGE.

Acknowledgment

This work was supported by the National Key Plan for Scientific Research and Development of China (Grant No. 2016YFD0500307).

References

1. Lu G et al (2013) Molecular basis of binding between novel human coronavirus MERS-CoV and its receptor CD26. Nature 500 (7461):227–231
2. Yu X et al (2015) Structural basis for the neutralization of MERS-CoV by a human monoclonal antibody MERS-27. Sci Rep 5:13133
3. Zhang S et al (2018) Structural definition of a unique neutralization epitope on the receptor-binding domain of MERS-CoV spike glycoprotein. Cell Rep 24(2):441–452
4. Niu P et al (2018) Ultrapotent human neutralizing antibody repertoires against Middle East respiratory syndrome coronavirus from a recovered patient. J Infect Dis 218 (8):1249–1260

5. Wang L et al (2015) Evaluation of candidate vaccine approaches for MERS-CoV. Nat Commun 6:7712
6. Ying T et al (2015) Junctional and allele-specific residues are critical for MERS-CoV neutralization by an exceptionally potent germline-like antibody. Nat Commun 6:8223
7. Du L et al (2016) Introduction of neutralizing immunogenicity index to the rational design of MERS coronavirus subunit vaccines. Nat Commun 7:13473
8. Wang L et al (2018) Importance of neutralizing monoclonal antibodies targeting multiple antigenic sites on MERS-CoV spike to avoid neutralization escape. J Virol 92(10). https://doi.org/10.1128/JVI.02002-17
9. Wang N et al (2013) Structure of MERS-CoV spike receptor-binding domain complexed with human receptor DPP4. Cell Res 23 (8):986–993
10. Otwinowski Z, Minor W (1997) Processing of X-ray diffraction data collected in oscillation mode. Methods Enzymol 276:307–326
11. McCoy AJ et al (2007) Phaser crystallographic software. J Appl Crystallogr 40(Pt 4):658–674
12. Emsley P, Cowtan K (2004) Coot: model-building tools for molecular graphics. Acta Crystallogr D Biol Crystallogr 60(Pt 12 Pt 1):2126–2132
13. Adams PD et al (2002) PHENIX: building new software for automated crystallographic structure determination. Acta Crystallogr D Biol Crystallogr 58(Pt 11):1948–1954

PART II

Genetic Alteration and Structural Determination of MERS-Coronavirus Proteins

Chapter 5

Bacterial Artificial Chromosome-Based Lambda Red Recombination with the I-SceI Homing Endonuclease for Genetic Alteration of MERS-CoV

Anthony R. Fehr

Abstract

Over the past two decades, several coronavirus (CoV) infectious clones have been engineered, allowing for the manipulation of their large viral genomes (~30 kb) using unique reverse genetic systems. These reverse genetic systems include targeted recombination, in vitro ligation, vaccinia virus vectors, and bacterial artificial chromosomes (BACs). Quickly after the identification of Middle East respiratory syndrome-CoV (MERS-CoV), both in vitro ligation and BAC-based reverse genetic technologies were engineered for MERS-CoV to study its basic biological properties, develop live-attenuated vaccines, and test antiviral drugs. Here, I will describe how lambda red recombination can be used with the MERS-CoV BAC to quickly and efficiently introduce virtually any type of genetic modification (point mutations, insertions, deletions) into the MERS-CoV genome and recover recombinant virus.

Key words Coronavirus, MERS-CoV, Bacterial artificial chromosome (BAC), Lambda red recombination, Reverse genetics, Infectious clone

1 Introduction

Coronaviruses are large, enveloped, single-stranded positive-sense RNA viruses that cause both significant human and veterinary disease. Prior to the severe acute respiratory syndrome-CoV (SARS-CoV) outbreak in 2003, human CoVs were only known to cause mild, self-limiting upper respiratory diseases. Approximately 10 years after the emergence of SARS-CoV in 2012, Middle East respiratory syndrome (MERS)-CoV emerged in the Middle East where it then spread to 27 different countries, and to date (December 2018, WHO) there have been 2278 laboratory-confirmed cases and 806 associated deaths for a case fatality rate of 35%. Most of these cases have occurred in the Middle East, aside from an outbreak of ~200 infected individuals in South Korea in 2015 [1].

Rahul Vijay (ed.), *MERS Coronavirus: Methods and Protocols*, Methods in Molecular Biology, vol. 2099,
https://doi.org/10.1007/978-1-0716-0211-9_5,

Infectious clones are highly valuable research tools that enable modification of viral genomes to better understand their fundamental biology, develop novel vaccine candidates, and test antiviral therapeutics. Soon after identifying MERS-CoV as the causative agent of MERS, two distinct infectious clones were reported for MERS-CoV [2, 3]. These infectious clones were engineered using in vitro ligation or bacterial artificial chromosomes (BACs), each of which had been used previously for CoVs [4–6]. In vitro ligation uses unique type II restriction endonucleases that cleave several bases away from their recognition site, allowing for the reassembly of authentic CoV genomes from smaller fragments. Each fragment is separately maintained in its own small plasmid for efficient genetic modification using traditional molecular cloning methods. Separating specific nucleotide sequences in ORF1A helped to eliminate the problem of these sequences being toxic for bacteria. A T7 promoter is inserted at the 5′ end of the genome, allowing for in vitro transcription of the viral RNA and subsequent transfection into mammalian cells for virus production. In contrast, BACs allow for the stable propagation of full-length CoV cDNA in bacteria, due to the ability to restrict their copy number to 1 or 2 plasmids per cell. Different restriction fragments of these BACs can be sub-cloned into smaller vectors for efficient modification, or the full-length genome can be modified using lambda red recombination, which will be discussed here. CoV BAC plasmids contain a CMV promoter 5′ of the viral genome, allowing for transcription of the viral genome following transfection of BAC DNA into mammalian cells. In addition, the CoV BACs contain a polyA tail, a Hepatitis D Virus (HDV) ribozyme, and bovine growth hormone (BGH) termination and polyadenylation signals to create genomic RNA with an authentic 3′ end. The full-length nature of BAC DNA and the CMV promoter subvert the need for in vitro ligation or transcription to recover infectious virus. BACs were initially developed in the early 1990s, and by the mid-late 1990s they were utilized by Herpes virologists for modification of these large DNA viruses, which revolutionized the field. A few years later a BAC for a CoV, transmissible gastroenteritis virus (TGEV), was engineered, and since then BACs have been successfully developed for several CoVs including feline infectious peritonitis virus (FIPV), OC-43, SARS-CoV, MERS-CoV, murine hepatitis virus strain JHM (MHV-JHM), porcine epidemic diarrhea virus (PEDV), and the SARS-like CoV WIV-1 [2, 4, 7–12]. Thus, it is likely that BAC-based reverse genetics could be useful for any novel or emergent CoV.

Lambda red recombination utilizes bacteriophage enzymes Exo, Beta, and Gam (Red proteins) to mediate homologous recombination near the ends of linear double-stranded DNA [13, 14]. PCR products containing positive selection markers are suitable substrates for these enzymes, so long as they bear extensions of 40–50 bases that are homologous to the target sequence. A major advancement in this technique came with the development of an *E. coli* strain, DY380, where the Red proteins were placed under the control of a temperature-inducible promoter [15]. Several methods for removing the positive selection markers from the viral genomes have been developed, including flanking sequences with FRT or loxP sites [16], or utilizing positive and negative selection markers on a single gene cassette, such as the Galactose Kinase (GalK)-Kan^R gene cassette [17]. These methods both have certain downfalls, including the retention of small FRT or loxP sites following removal of the marker, or the unintended removal of negative selection markers by repeat sequences in the BAC plasmid. To improve the efficiency of removing the positive selection marker, a unique method utilizing the I-SceI homing endonuclease under an arabinose-inducible promoter was developed (Fig. 1) [18]. I-SceI is an endonuclease with an 18 bp recognition site that is not present in the *E. coli* genome, making it safe to express in *E. coli*. In the method described here, this recognition site is engineered on a plasmid (pEP-KanS) just outside of the positive selection marker, and its cleavage with the I-SceI enzyme allows for the removal of the positive selection marker by intramolecular Red recombination utilizing sequence duplication introduced in the original PCR primers. This method can be utilized to introduce any type of modification into the BAC DNA, including mutations, deletions, and insertions. Here I will outline the procedure for this highly efficient method to engineer markerless modifications, focusing on single point mutations in the full-length MERS-CoV BAC.

2 Materials

2.1 Manipulation of the MERS-CoV BAC

2.1.1 Plasmids and Bacterial Strains

1. pBAC-MERS-CoV^{FL}. This MERS-CoV BAC was first engineered by the Luis Enjuanes lab [2]. The full protocol for creating this BAC was subsequently published by the same group in a previous Methods in Molecular Biology book [19]. This plasmid contains the parA, parB, and parC genes derived from the *E. coli* F-factor to prevent more than one or two BACs from coexisting in the same cell. It also contains genes involved in the initiation and orientation of DNA replication and the chloramphenicol resistance gene (*Cml^r^*).
2. pEP-KanS. This plasmid contains the AphAI-I-SceI cassette containing a kanamycin resistance marker (*Kan^r^*) and an I-SceI restriction site [18]. This plasmid also contains an ampicillin resistance marker.
3. *E. coli* strains DH10B (*see* **Note1**) and GS1783 (*see* **Note 2**) cells.

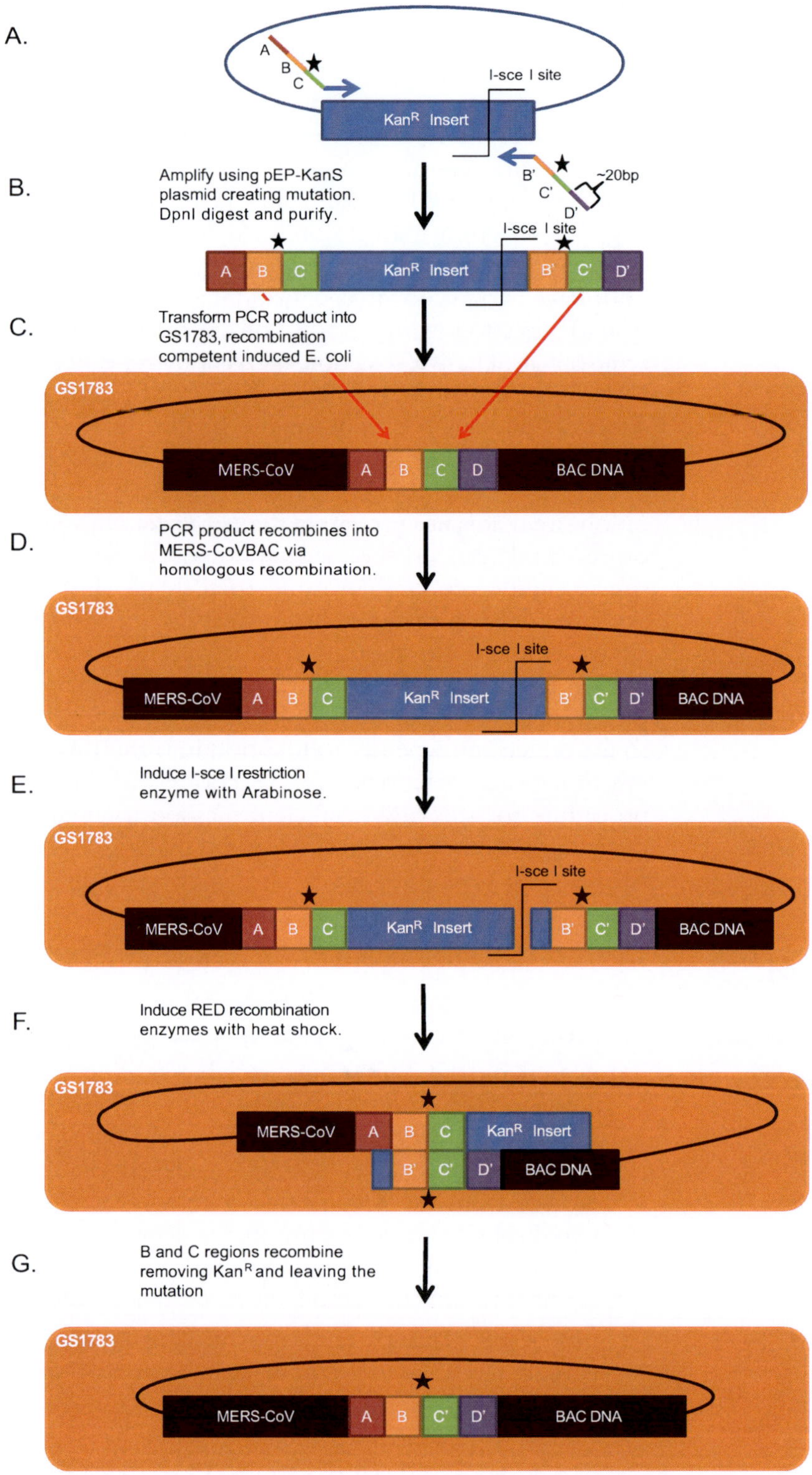

Fig. 1 Schematic for making point mutations in MERS-CoV using Lambda Red Recombination. This diagram illustrates and describes each individual step (**a**–**g**) in the protocol for creating individual point mutations in the MERS-CoV BAC

2.1.2 Culture and Freezing Reagents for E. coli

1. LB medium: 1% (w/v) tryptone, 0.5% (w/v) yeast extract, 1% (w/v) NaCl. Sterilize by autoclaving on liquid cycle.
2. LB agar plates: LB medium containing 15 g/L agar. After preparing LB medium add the agar. Sterilize by autoclaving as above. Allow the medium to cool to ~45–50 °C, add appropriate antibiotics (1 mL of 1000× stocks/L) or arabinose (40 mL of 25% arabinose/L) to the medium, then dispense in Petri dishes.
3. LB freezing medium: 40% (v/v) glycerol in LB medium. Add glycerol, water, and dry LB ingredients to desired volume (i.e., 200 mL of glycerol for 500 mL total LB freezing medium). Sterilize by autoclaving on liquid cycle.
4. LB cml media: LB medium with chloramphenicol (25 mg/mL).
5. LB cml/kan media: LB media with chloramphenicol (25 mg/mL) and kanamycin (40 mg/mL).
6. SOC medium: 2% (w/v) tryptone, 0.5% (w/v) yeast extract, 0.05% (w/v) NaCl, 2.5 mM KCl, 10 mM MgCl2, 10 mM MgSO4, 20 mM glucose. Sterilize by autoclaving on liquid cycle.
7. Antibiotics: Make 1000× stock solutions of ampicillin (100 mg/mL in ethanol), kanamycin (40 mg/mL in H_2O), and chloramphenicol (25 mg/mL in ethanol).
8. Arabinose: Make 25% (w/v) solution in H_2O. Sterilize by passing it through a 0.22 μm disposable filter.
9. Glycerol: Make a 10% (v/v) solution and sterilize by autoclaving on liquid cycle.

2.1.3 Enzymes

1. Restriction endonucleases, Taq DNA polymerase, high-fidelity thermostable DNA polymerase, and reverse transcriptase can be purchased from several different commercial sources.

2.1.4 DNA Oligomers

1. DNA oligomers can be purchased from several different commercial sources. Long DNA oligos (>80 nt) are needed, so identifying a commercial source that can make long oligos at a reasonable price is important (*see* **Note 3**).
2. Recombination Primers.

   Forward: 5′----------60bp_homology------AGGATGACGACGATAAGTAGGG-3′.

   Reverse: 5′----------60bp_homology------GCCAGTGTTACAACCAATTAACC-3′.

2.1.5 DNA Preparation Kits

1. DNA Miniprep Kit.
2. Machery-Nagel Nucleobond Xtra Midi Kit (*see* **Note 4**).
3. Invitrogen PureLink PCR Purification Kit (*see* **Note 5**).

2.1.6 Special Software and Equipment

1. Electroporator and 1 mm cuvettes.
2. 42 °C shaking water bath (*see* **Note 6**).
3. 30–32 °C incubator and shaking incubator.
4. DNA analysis software.

2.2 Rescue of BAC-Derived Recombinant Viruses

1. Human liver-derived Huh-7 (JCRB Cell Bank, JCRB0403) or Vero cells (ATCC, CCL-81).
2. Cell growth medium: Dulbecco's Modified Eagle Medium (DMEM) + 10% FBS.
3. Cell growth medium: DMEM + 2% FBS.
4. Opti-MEM I Reduced Serum Medium.
5. Trypsin-EDTA: 0.05% (w/v) trypsin, 0.02% (w/v) EDTA.
6. Lipofectamine 2000 (Life Technologies, Invitrogen) (*see* **Note 7**).

3 Methods

3.1 Transformation and Storage of MERS-CoV BAC DNA into DH10B or GS1783 Cells

1. Precool electroporation cuvette at 4 °C or on ice.
2. Prepare labeled 14 mL culture tube(s) with 1 mL SOC.
3. Add 1–10 ng of BAC DNA in a sterile Eppendorf tube on ice. Add 24 μL of electrocompetent DH10B or GS1783 *E. coli*.
4. Carefully transfer the mixture into the groove of an electroporation cuvette on ice.
5. Pulse the cuvette at 25 μF, 1750 V, and 200 Ω.
6. Recover by adding 0.5–1.0 mL SOC to the cuvette, and transfer the mixture to a 14 mL culture tube. Incubate at 32 °C and 220 rpm for 1 h. Pre-warm LB-agar-cml plate.
7. Add 25–50 μL of culture to LB-cml plate. Incubate at 30–32 °C overnight (o/n) (*see* **Note 8**).
8. Next day pick 1–2 colonies and incubate in 2 mL LB-cml broth at 30–32 °C o/n.
9. Next day mix 0.5 mL o/n culture with 0.5 mL bacterial freezing medium. Store in negative 80 °C freezer.

3.2 Prepare pBAC-MERS-CoV Competent Cells for Lambda Red Recombination

Day 1

1. Streak glycerol stock of GS1783 pBAC-MERS-CoV bacteria generated in 3.1 on LB-cml plate to generate isolated colonies. Incubate at 30–32 °C o/n.

Day 2

2. Pick one colony from the plate and transfer to 2 mL LB-cml in tube. Incubate at 30–32 °C and 220 rpm o/n. Put ddH_2O (200 mL) and 10% glycerol in ddH_2O (100 mL) at 4 °C to precool.

Day 3

3. Dilute 1 mL of o/n culture into 49 mL LB-cml in a 250 mL Erlenmeyer flask. Protocol can be scaled up to 100 mL (use 500 mL Erlenmeyer flask). Incubate at 32 °C and 220 rpm. In the meantime, turn on shaking water bath, check water level, and set to 42 °C. Label 10 Eppendorf tubes per 50 mL culture and precool them in a plastic tube rack at −80 °C. Precool the tabletop centrifuge with the swinging bucket rotor and a microcentrifuge to 4 °C.
4. Monitor the OD_{600} of culture starting at 2 h of incubation. When OD_{600} = 0.6–0.8 (~3–4 h), incubate at 42 °C and 200 rpm (*see* **Note 9**) in shaking water bath for *15 min*. In the meantime, prepare an ice water slurry in an autoclave bin or other suitable container and get a bucket of ice. Place two 50 mL conical tubes on ice.
5. Swirl the Erlenmeyer flask in the ice water slurry for *10 min*. From here on, keep bacteria cold at all times.
6. Aliquot 25 mL into each 50 mL conical tube on ice. Centrifuge at 1800 × *g* for 10 min at 4 °C in a tabletop centrifuge.
7. Pour off supernatant with one quick motion. While pouring, position the bacterial pellet away from the liquid to limit the amount of bacteria lost. Add 5 mL ice-cold sterile ddH_2O and resuspend pellet by swirling and tapping the tube to the bottom of the ice-cold autoclave bin. Once resuspended, add an additional 20 mL ice-cold sterile ddH_2O and centrifuge at 1800 × *g* for 10 min at 4 °C. Repeat 1×.
8. Following the second water wash, resuspend each pellet with 10% glycerol, first in 5 mL, then add an additional 15 mL. Centrifuge at 1800 × g for 10 min at 4 °C.
9. Pour off supernatant as described above. Following the pour, resuspend the pellet in the remaining 10% glycerol (~500 μL). If the combined amount of cells and 10% glycerol is greater than 550 μL, transfer the suspension to a cold Eppendorf tube and pellet the cells for 2 min at 5000 × g and 4 °C. After spinning the cells, remove an appropriate amount of supernatant such that ~500 μL of cell suspension remains. Resuspend the cells to a homogenous solution and aliquot 50 μL to prefrozen Eppendorf tubes and flash-freeze tubes in liquid nitrogen or a dry ice-methanol bath. Use immediately or store at −80 °C. Cells are typically good for 6–12 months, but may be useful even after several years.

3.3 Create PCR Cassette Containing the Viral Genome Mutation or Insertion of Interest with the AphAI-I-SceI (Herein Termed Kan^r-I-SceI) Gene Cassette

1. Design and order *Kan^r*-I-SceI primers with 60 bp homology to the region of interest flanking the desired site to be modified (Fig. 1a). To design simple point mutations, start by developing a 60 bp flanking sequence, calling each 20 base pair section as A, B, and C. Then incorporate the desired mutation at the end of section B (40th base pair, or the 39th and 40th base pair if two changes are required). Finally, attach this sequence to the 22 bp *Kan^r*-I-SceI sequence below to create your forward recombination primer. To create the reverse recombination primer, create a new block of 20 bp we will call section D′ that is the reverse complement of the sequence immediately downstream from section C. These 20 bp will be followed on your primer by sections C′ and B′, the reverse complements to sections B and C, with B′ containing the desired mutations. Finally, add the 23 bp *Kan^r*-I-SceI sequence to finish the reverse primer. During negative selection, sections B/C will recombine with B′/C′ leading to the loss of the *Kan^R*-I-SceI cassette (Fig. 1g). For deletion mutants, leave out the desired sequence from your primers. For instance, to delete sections D/E/F, simply create the forward primer with sections A/B/C, and the reverse primer with sections G′/C′/B′. Insertions of small sequences can be achieved by adding the entire insertion sequence at the 3′ end of the forward primer, and at least 50 bp of reverse complement sequence at the 3′ end of the reverse primer (Fig. 2). Larger insertions may require the development of a full plasmid, or potentially the use of nested PCRs. For additional details, *see* ref. [18]. While designing recombination primers (indicated below), remember to also order short primers about 100–200 bp outside of the insertion site to check for the proper insertion of the gene cassette by PCR.
2. Set up PCR reaction and perform reaction according to manufacturer's protocol with following modifications (Fig. 1b).

Fig. 2 Model of the PCR product used for inserting specific sequences into BACs using lambda red recombination. The full sequence for insertion is incorporated at the 5′ end of the *Kan^R*-I-SceI cassette while at least 50 nt of sequence homologous to the 3′ end of the insertion sequence is incorporated at the 3′ end of this cassette. Surrounding these sequences are 50 nt of sequence homologous to the viral sequence where the sequence is to be inserted

3. For a PCR template, use ~50 ng of the pEP-KanS plasmid.
4. Use 1 μL high-fidelity polymerase (*see* **Note 10**).
5. For the annealing temperature, I prefer to use a step-down procedure lowering the annealing temperature by 1 °C starting at 68 °C and continuing the PCR reaction for 25 cycles. Using this method we rarely see spurious PCR products.
6. Analyze the PCR product on an agarose gel with ethidium bromide and image it using a UV-gel box and gel-imaging software. PCR product should be ~1.2 kb.
7. Purify the PCR product using the PureLink PCR purification kit (*see* **Note 5**). Use binding buffer B3 according to manufacturer's protocol to remove primer dimers from the mixture. Elute DNA into 44 μL of water.
8. DpnI digest the pEP-KanS plasmid in the purified PCR product (Fig. 1b). DpnI specifically cleaves methylated DNA and is needed to digest the pEP-KanS plasmid used in the PCR reaction. It has a 4 bp recognition site so it should cleave DNA approximately every 250 bp. Without this digestion all of your transformants will maintain the pEP-KanS plasmid as its transformation is much more efficient than the recombination of the PCR product. Incubate for 1–3 h at 37 °C.

   1 μL DpnI.

   5 μL Buffer.

   44 μL PCR product.

   50 μL Total

9. Purify PCR product using the PureLink PCR purification kit. Elute DNA into 30 μL pre-warmed elution buffer or water.
10. Measure DNA concentration using a spectrophotometer. A concentration of 20–80 ng/μL is typical.

3.4 Insertion of Kan^R-I-SceI Gene Cassette into MERS-CoV BAC DNA (Positive Selection)

Day 1

1. Precool the electroporation cuvette(s) on ice.
2. Prepare labeled 14 mL culture tube(s).
3. Aliquot 50–100 ng of *Kan^r*-I-SceI PCR product in a sterile Eppendorf tube on ice. This will generally be 2–4 μL. Add 23 μL competent GS1783 *E. coli* containing pBAC-MERS-CoV and mix by stirring briefly.
4. Carefully transfer the mixture into the groove of the electroporation cuvette on ice.
5. Wipe any ice water from outside of cuvette and pulse at 25 μF, 1750 V, and 200 Ω (Fig. 1c).

6. Recover by immediately adding 0.5–1 mL SOC to the cuvette, and then transfer the mixture to a 14 mL culture tube. Incubate at 32 °C and 220 rpm for 3–5 h. This is when the recombination occurs (Fig. 1d) (*see* **Note 11**). Pre-warm an LB-cml/kan plate.
7. After the recovery, transfer the bacteria to a 1.5 mL tube and centrifuge at 16,000 × g for 1 min. Remove all but ~100 μL of supernatant, resuspend the pellet, and plate the entire culture on an LB-cml/kan plate and incubate at 30–32 °C o/n. You should get anywhere from 5 to 100 colonies.

Day 2

1. Using a sterile toothpick or pipet tip, replica-plate 25 colonies from the previous step first onto an LB-amp plate (ampR = pEP-kanS plasmid background; all colonies should be negative if DpnI digest was complete), then onto an LB-cml/kan plate (should be positive if recombination occurred). A grid for this procedure is shown in Fig. 3. Incubate plates at 30–32 °C o/n.

Day 3

2. Identify bacterial clones that grew on the LB-cml/kan plate but not on the LB-amp plate. The efficiency at this step is anywhere from 50 to 100%. Inoculate culture tubes containing 6 mL of LB-cml-kan with 5–6 selected colonies (1 in each tube) from the LB-cml/kan plate and incubate at 30–32 °C and 220 rpm o/n.

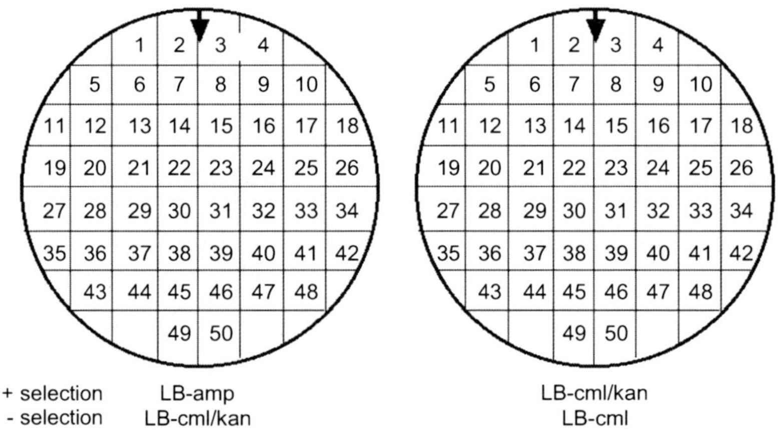

Fig. 3 Replica plate grids. These grids allow for the easy identification of identical colonies that have been plated on each plate. Using a toothpick, dot a single colony in the same spot on each plate. For both positive and negative selection, MERS-BAC clones that have successfully undergone recombination will grow on the plate on the right, but not on the plate on the left

Day 4

3. Following o/n incubation, create a freezer stock of the bacteria, then purify the BAC DNA using a standard miniprep kit. Using 1 μL of the BAC DNA, use the external primers located 100–200 bp outside the region of homology you previously designed to test for the insertion of *Kan*^{r}-I-SceI by PCR. If the insertion was successful, the DNA band from the PCR should be ~1 kb larger than the band from the MERS-CoV wild-type BAC (*see* **Note 12**). To speed this process up, colonies may be collected off the plate on day 3 and directly tested by PCR. Colonies that pass the PCR screen can then progress to the negative selection protocol (3.5) (*see* **Note 13**).

3.5 Removal of Kan^R-I-SceI from MERS-CoV BAC DNA (Negative Selection)

Day 1 (can coincide with day 4 of positive selection protocol)

1. Create two new culture tubes with 2 mL of LB-cml/kan and inoculate them with bacteria from 2 separate *kan*^{r}+*cml*^{r}+*amp*^{r}− colonies. Incubate at 30–32 °C and 220 rpm o/n.

Day 2

2. Transfer 100 μL of the fresh o/n culture to 2 mL of warm LB-cml and incubate at 32 °C and 220 rpm for 2 h. If not previously done, mix 0.5 mL o/n culture with 0.5 mL freezing medium as a glycerol stock.
3. After 2 h, add 2 mL of warm LB-cml with 2% arabinose to the culture tube for a final arabinose concentration of 1% (Fig. 1). Incubate at 32 °C and 220 rpm for 2 h. Warm water bath shaker to 42 °C.
4. Transfer the culture tubes to the water bath shaker at 42 °C and ~200 rpm and incubate for 30 min (Fig. 1f).
5. Transfer the culture tubes back to 32 °C and incubate for 3–4 h (Fig. 1g).
6. After the incubation, perform tenfold serial dilutions of the bacteria in LB. Plate 100 μL of 10^{-4} and 10^{-5} dilutions of original culture on pre-warmed LB-cml plates containing 1% arabinose. Incubate the plates at 30–32 °C.

Day 3

7. Pick 50 colonies and replica plate on LB-cml/kan and LB-cml plates. Colonies that underwent correct recombination should grow on LB-cml but not on LB-cml/kan. Incubate o/n at 32 °C. Efficiency is generally anywhere from 5 to 50% (*see* **Note 14**).

Day 4

8. Pick 3 separate cml$^+$ kan$^-$ colonies and culture each one in 100 mL LB-cml at 32 °C and 220 rpm o/n.

Day 5

9. Mix 0.5 mL of the o/n culture with 0.5 mL freezing solution for a glycerol stock. Purify the BAC DNA from these cultures using the NucleoBond Xtra Midi Kit according the manufacturer's protocol (*see* **Note 4**). Verify the integrity of the BAC constructs by restriction enzyme digestion of ~2 μg BAC DNA with KpnI and verify the loss of the Kan^R-I-SceI gene cassette by PCR (*see* **Note 15**). Verify introduced mutation(s)/insertion(s) by Sanger sequencing.

3.6 Rescue of Recombinant Virus

Day 1

1. Seed either Vero-81 or Huh-7 cells (*see* **Note 16**) in DMEM + 10% FBS into 6 well dish so that cells are 60–80% confluent at the time of transfection the next day. Incubate at 37 °C o/n.

Day 2 (BSL-3)

2. All procedures from here involve working with MERS-CoV which requires a BSL-3 containment laboratory.
3. Replace medium from cells with 2 mL DMEM +10% FBS immediately before proceeding to **step4**.
4. Prepare the transfection mixture as follows: For a single well, mix 1–2 μg of MERS-CoV BAC DNA with 10 μL Lipofectamine 2000 (*see* **Note** 7) in Opti-MEM media according to manufacturer's protocol. Scale up accordingly if multiple BAC DNAs will be transfected. Additionally, prepare the following controls:

   (a) Lipofectamine + random plasmid DNA.

   (b) Lipofectamine only.

   (c) No Lipofectamine, no DNA.

   Wait ~20 min for liposomes to form. Then add the transfection mixture dropwise to the medium in the well.

5. Approximately 3–4 h after transfection (*see* **Note 17**), replace the medium with a 2% (Vero 81 cells) or 10% (Huh-7 cells) FBS medium to slow cell growth.

Day 6 and Beyond (BSL-3)

6. Cytopathic effect (CPE) will start being visible at 3–4 days after transfection. Collect cells and supernatant when >50% of well has CPE (the more the better). To collect, scrape any

remaining cells off the well with a pipette tip or cell scraper and transfer the media and cell debris in a 2 mL microcentrifuge tube and freeze-thaw the sample. Then, centrifuge the sample at ~5000 × g to spin out the cell debris and transfer the supernatant to a new tube. We term this passage 0 (P0) virus.

7. Use ~1–2 mL of P0 virus to infect 1 T-75 flask of cells (*see* **Note 18**).
8. Collect virus at 2–3 days post-infection as describe above and aliquot. We term this as P1 virus. Titer virus on Vero/Huh-7 cells. If titer is sufficient, use for further experiments.
9. To verify whether the virus has maintained your mutation or insertion, infect a new set of cells for ~48 h, then collect the infected cells in Trizol, prepare RNA and then convert the RNA to cDNA with reverse transcriptase. Alternatively, you could prepare RNA from the supernatants and sequence these samples. Create a PCR product with the external primers that were previously used to check for the gain or loss of the insertion, and finally have that PCR product sequenced by Sanger sequencing (*see* **Note 19**).

4 Notes

1. DH10B cells are a standard *E. coli* strain used for general cloning. DH10B cells are recombination-deficient (recA1) and endonuclease I-deficient (endA1) and have constitutive deoxyribose synthesis for improved cloning of large plasmids. These traits make these cells ideal for the long-term storage of BAC plasmids.
2. GS1783 cells are a derivative of DH10B cells that contain the Red recombination genes under a temperature sensitive promoter and the I-SceI homing endonuclease under control of an arabinose-inducible promoter [20].
3. For creating long oligos, we prefer Invitrogen as a supplier, as they can provide oligos up to 100 nt at their standard price per base. Many companies do not make oligos longer than 60 nt at their standard rates, and thus will charge a significant amount more for the 80–90 nt oligos required for this protocol.
4. While many companies sell BAC-prep kits which will work for these purposes, these columns are often at least 3× the cost of the Nucleobond Xtra Midi Kit. We have tested the Xtra Midi Kit side by side with a BAC-prep kit and found little to no difference in CoV BAC DNA yield.
5. For purifying PCR products, we prefer the Invitrogen PureLink PCR Purification Kit because it includes a buffer that allows DNA products of <300 bp to go through the column.

This helps remove primer dimers from the PCR reaction that could interfere with recombination.

6. A 42 °C shaking water bath is essential for this procedure. The bacterial cultures need to heat up to 42 °C quickly to properly induce the Red enzymes. A shaking water bath is significantly better than a regular shaking incubator at quickly transferring heat to the bacterial culture.
7. We have tried several different transfection reagents, and Lipofectamine 2000 has worked the best for us. That does not mean other transfection reagents won't work, feel free to try whichever reagent you prefer.
8. It is important to always culture GS1783 cells at 32 °C or lower due to the potential for leaky expression of the Red enzymes at temperatures below the induction temperature of 42 °C.
9. Some shaking water baths cannot shake at 200 rpm; if this is the case, simply shake at a reasonable speed for the shaker.
10. We have occasionally had problems obtaining a PCR product when using the manufacturer recommended 0.5 μL of high-fidelity polymerase. Using 1 μL of polymerase provides more consistent results.
11. Unlike a typical 1 h recovery following electroporation, it is important to incubate these cultures for several hours in SOC to allow time for recombination to occur. A recovery time of 5 h or greater is preferred, with a minimal recovery time being 3 h.
12. Occasionally, we find that following PCR some clones have bands that correspond to both the WT BAC and the desired Kan^R-I-SceI insert BAC. It is likely due to at least two copies of BAC DNA being present in the same cell. Unless the molar ratio of the Kan^R-I-SceI insert to the WT BAC is very large, these clones should be avoided.
13. After positive selection, we do not check the BAC by restriction digest, because (a) we find that in many cases the amount of BAC DNA from a miniprep is insufficient for a readable digest, and (b) we find that the colonies that pass the replica plating and PCR tests rarely if ever have any significant problems concerning removing or duplicating regions of the MERS-CoV BAC DNA.
14. First, colonies may take over 1 day before they are visible. Second, while the efficiency of the negative selection can be very low, any colonies that have grown on LB-cml plates but do not grow on LB-cml/kan plates are very likely correct. Therefore, we go directly to starting large-scale cultures for these clones. That way we can do the final diagnostic tests and we can prepare for the BAC transfection and viral recovery at the same time.

15. A KpnI digest of pBAC-MERS-CoV results in DNA fragments of 19.1, 13.8, 3.9, and 1.5 kb. We have found this is the best digest for diagnostic evaluation of the MERS-CoV BAC; however other enzymes or enzyme combinations will work as well. Be sure to use a low-percentage agarose gel to effectively separate the large DNA molecules.
16. Other laboratories use BHK-21 cells for the initial transfection since these cells are highly transfectable. The transfected cells can be overlaid on the Vero-81 or Huh-7 cells for further outgrowth of the recombinant virus.
17. It is feasible to wait o/n to change the media of the transfected Vero 81 cells, but this does result in an increase in the cytotoxicity induced by Lipofectamine. This is not feasible if you are using Huh-7 cells.
18. Mutant viruses that do not replicate as well as WT virus may require increased amounts of P0 virus.
19. It is also beneficial to check the integrity of the entire MERS-CoV genome by RT-PCR after several passages as MERS-CoV tends to occasionally delete sections of the accessory proteins.

Acknowledgments

I thank Jeremiah Athmer and Andrea Pruijssers for their substantial modifications and improvements to this protocol over the years. I also thank Catherine Kerr, Ethan Doerger, Andrea Pruijssers, Isabel Sola, and Luis Enjuanes for critical reading of this manuscript. This work was supported in part by NIH grants CoBRE P20 GM113117-02 and K22 AI134993-01, and start-up funds from the University of Kansas. The funders had no role in study design, data collection and interpretation, or the decision to submit the work for publication.

References

1. Fehr AR, Channappanavar R, Perlman S (2017) Middle East respiratory syndrome: emergence of a pathogenic human coronavirus. Annu Rev Med 68:387–399. https://doi.org/10.1146/annurev-med-051215-031152
2. Almazan F, DeDiego ML, Sola I, Zuniga S, Nieto-Torres JL, Marquez-Jurado S, Andres G, Enjuanes L (2013) Engineering a replication-competent, propagation-defective Middle East respiratory syndrome coronavirus as a vaccine candidate. MBio 4(5): e00650–e00613. https://doi.org/10.1128/mBio.00650-13
3. Scobey T, Yount BL, Sims AC, Donaldson EF, Agnihothram SS, Menachery VD, Graham RL, Swanstrom J, Bove PF, Kim JD, Grego S, Randell SH, Baric RS (2013) Reverse genetics with a full-length infectious cDNA of the Middle East respiratory syndrome coronavirus. Proc Natl Acad Sci U S A 110(40):16157–16162. https://doi.org/10.1073/pnas.1311542110
4. Almazan F, Gonzalez JM, Penzes Z, Izeta A, Calvo E, Plana-Duran J, Enjuanes L (2000) Engineering the largest RNA virus genome as an infectious bacterial artificial chromosome. Proc Natl Acad Sci U S A 97(10):5516–5521

5. Yount B, Denison MR, Weiss SR, Baric RS (2002) Systematic assembly of a full-length infectious cDNA of mouse hepatitis virus strain A59. J Virol 76(21):11065–11078
6. Yount B, Curtis KM, Baric RS (2000) Strategy for systematic assembly of large RNA and DNA genomes: transmissible gastroenteritis virus model. J Virol 74(22):10600–10611
7. Almazan F, Dediego ML, Galan C, Escors D, Alvarez E, Ortego J, Sola I, Zuniga S, Alonso S, Moreno JL, Nogales A, Capiscol C, Enjuanes L (2006) Construction of a severe acute respiratory syndrome coronavirus infectious cDNA clone and a replicon to study coronavirus RNA synthesis. J Virol 80 (21):10900–10906. https://doi.org/10.1128/JVI.00385-06
8. Balint A, Farsang A, Zadori Z, Hornyak A, Dencso L, Almazan F, Enjuanes L, Belak S (2012) Molecular characterization of feline infectious peritonitis virus strain DF-2 and studies of the role of ORF3abc in viral cell tropism. J Virol 86(11):6258–6267. https://doi.org/10.1128/JVI.00189-12
9. Fehr AR, Athmer J, Channappanavar R, Phillips JM, Meyerholz DK, Perlman S (2015) The nsp3 macrodomain promotes virulence in mice with coronavirus-induced encephalitis. J Virol 89(3):1523–1536. https://doi.org/10.1128/JVI.02596-14
10. St-Jean JR, Desforges M, Almazan F, Jacomy H, Enjuanes L, Talbot PJ (2006) Recovery of a neurovirulent human coronavirus OC43 from an infectious cDNA clone. J Virol 80(7):3670–3674. https://doi.org/10.1128/JVI.80.7.3670-3674.2006
11. Li J, Jin Z, Gao Y, Zhou L, Ge X, Guo X, Han J, Yang H (2017) Development of the full-length cDNA clones of two porcine epidemic diarrhea disease virus isolates with different virulence. PLoS One 12(3):e0173998. https://doi.org/10.1371/journal.pone.0173998
12. Zeng LP, Gao YT, Ge XY, Zhang Q, Peng C, Yang XL, Tan B, Chen J, Chmura AA, Daszak P, Shi ZL (2016) Bat severe acute respiratory syndrome-like coronavirus WIV1 encodes an extra accessory protein, ORFX, involved in modulation of the host immune response. J Virol 90(14):6573–6582. https://doi.org/10.1128/JVI.03079-15
13. Murphy KC (1998) Use of bacteriophage lambda recombination functions to promote gene replacement in Escherichia coli. J Bacteriol 180(8):2063–2071
14. Oppenheim AB, Rattray AJ, Bubunenko M, Thomason LC, Court DL (2004) In vivo recombineering of bacteriophage lambda by PCR fragments and single-strand oligonucleotides. Virology 319(2):185–189. https://doi.org/10.1016/j.virol.2003.11.007
15. Lee EC, Yu D, Martinez de Velasco J, Tessarollo L, Swing DA, Court DL, Jenkins NA, Copeland NG (2001) A highly efficient Escherichia coli-based chromosome engineering system adapted for recombinogenic targeting and subcloning of BAC DNA. Genomics 73(1):56–65. https://doi.org/10.1006/geno.2000.6451
16. Zhang Y, Buchholz F, Muyrers JP, Stewart AF (1998) A new logic for DNA engineering using recombination in Escherichia coli. Nat Genet 20(2):123–128. https://doi.org/10.1038/2417
17. Warming S, Costantino N, Court DL, Jenkins NA, Copeland NG (2005) Simple and highly efficient BAC recombineering using galK selection. Nucleic Acids Res 33(4):e36. https://doi.org/10.1093/nar/gni035
18. Tischer BK, von Einem J, Kaufer B, Osterrieder N (2006) Two-step red-mediated recombination for versatile high-efficiency markerless DNA manipulation in Escherichia coli. BioTechniques 40(2):191–197
19. Almazan F, Marquez-Jurado S, Nogales A, Enjuanes L (2015) Engineering infectious cDNAs of coronavirus as bacterial artificial chromosomes. Methods Mol Biol 1282:135–152. https://doi.org/10.1007/978-1-4939-2438-7_13
20. Tischer BK, Smith GA, Osterrieder N (2010) En passant mutagenesis: a two step markerless red recombination system. Methods Mol Biol 634:421–430. https://doi.org/10.1007/978-1-60761-652-8_30

Chapter 6

Deducing the Crystal Structure of MERS-CoV Helicase

Sheng Cui and Wei Hao

Abstract

RNA virus encodes a helicase essential for viral RNA transcription and replication when the genome size is larger than 7 kb. Coronavirus (CoV) has an exceptionally large RNA genome (~30 kb) and it encodes an essential replicase, the nonstructural protein 13 (nsp13), a member of superfamily 1 helicases. Nsp13 is among the evolutionary most conserved proteins not only in CoVs but also in nidovirales. Thus, it is considered as an important drug target. However, the high-resolution structure of CoV nsp13 remained unavailable even until more than a decade after the outbreak of the severe acute respiratory syndrome coronavirus (SARS-CoV) in 2003, which hindered the structure-based drug design. This is in part due to the intrinsic flexibility of nsp13. Here, we describe protocols of deducing the crystal structure of Middle East respiratory syndrome coronavirus (MERS-CoV) helicase in detail, which include protein expression, purification, crystallization, enzymatic characterization, and structure determination. With these methods, catalytically active recombinant MERS-CoV nsp13 protein can be prepared and crystallized and the crystal structure can be solved.

Key words Coronavirus, Helicase, nsp13, Crystallization, Structure determination

1 Introduction

Coronavirus (CoV) remains a public health concern 16 years after the outbreak of the severe acute respiratory syndrome coronavirus (SARS-CoV) in 2003 [1]. The Middle East respiratory syndrome coronavirus (MERS-CoV) emerged in 2012, reemerged in 2015, and is still circulating in the Middle East region, which reminds the international community that the threat of CoVs persists [2, 3]. However, neither vaccine nor drugs against CoVs are currently available. Outbreaks of CoVs initiated extensive structural investigation on CoV encoded proteins thereafter, which not only shed light on the life cycle of CoVs but also laid foundation for the structure-based drug design (SBDD). CoV contains a positive single-stranded RNA genome of ~30 kb, one of the largest among +RNA viruses [4, 5]. To maintain the unusually large RNA genome, CoV encodes two replicase polyproteins pp1a and pp1ab, which are broken down into 16 nonstructural proteins

Rahul Vijay (ed.), *MERS Coronavirus: Methods and Protocols*, Methods in Molecular Biology, vol. 2099,
https://doi.org/10.1007/978-1-0716-0211-9_6,

(nsps) via proteinase cleavage [6, 7]. The nsps are then recruited to cytoplasm membranes, on which they form the membrane-associated replication-transcription complex (RTC). An RNA-dependent RNA polymerase nsp12 and a helicase nsp13 are the central components of RTC [8, 9]. However, while high-resolution structures of most CoV encoded proteins had been determined soon after SARS-CoV outbreak, the first CoV nsp13 structure, MERS-CoV nsp13, was only solved recently [10]. Nsp13 belongs to helicase superfamily 1 and shares conserved features with the eukaryotic Upfl helicase [11, 12]. Nsp13 is a multi-domain protein comprising of an N-terminal Cys/His rich domain (CH domain) and a C-terminal SF1 helicase core [10]. Nsp13 exhibits multiple enzymatic activities, including hydrolysis of NTPs and dNTPs, unwinding of DNA and RNA duplexes with 5′-3′ directionality and the RNA 5′-triphosphatase activity [13, 14]. To investigate the structure of CoV nsp13, we overexpressed the full-length MERS-CoV nsp13 (1-598aa) in insect cells and purified. The activity of the recombinant MERS-CoV nsp13 was verified by ATPase and helicase assays. Crystallization of MERS-CoV nsp13 was achieved by adding a synthetic single-stranded 15 poly dT DNA with 5′-triphosphate (ppp-15 T) to the protein, which restrains the intrinsic flexibility of nsp13. Benefiting from the presence of an N-terminal zinc-binding domain with three zinc atoms, multi-wavelength anomalous diffraction (MAD) data at the zinc absorption edge was collected, which allowed the determination of the crystal structure of MERS-CoV nsp13 [10].

2 Materials

Prepare all solutions using ultrapure water (prepared by purifying deionized water, to attain a sensitivity of 18 MΩ-cm at 25 °C) and analytical grade reagents. Prepare and store all reagents at room temperature (unless indicated otherwise). Diligently follow all waste disposal regulations when disposing waste materials. We do not add sodium azide to reagents.

2.1 Gene Cloning

1. Full-length MERS-CoV(GenBank accession: YP_009047202) nsp13 gene cDNA (GenScript).
2. The forward primer (gaaatt*ggatcc*gctgtcggttcatgc) and the reverse primer (gaaatt*ctcgag*tcactggagcttgtaatt) of full-length nsp13. Primers stocks are either supplied or diluted by molecular biology grade water to 100 μM and stored at −20 °C.
3. The pFastbac-1 baculovirus transfer vector is modified; 6 × Histidine-SUMO tag with a C terminal PreScission protease (PPase) site coding sequence in the N terminal of open reading frame [15].

4. Chemically competent bacterial cells of *E. coli* BL21 and *E. coli* DH10 Bac are prepared in-house as described.
5. High-fidelity PCR master mix with HF buffer (2 × fusion).
6. Endonuclease *Bam*HI and *Xho*I (fast digest).
7. 50× TAE buffer (1 L): 242 g Tris base, 57.1 mL glacial acetic acid and 100 mL of 0.5 M EDTA, pH 8.0, pH adjusted to 8.5. Filtered through a 0.2 μm membrane filter and used as a 1× solution.
8. Rapid DNA ligation kit (Promega).
9. LB medium: 5 g yeast extract, 10 g tryptone, and 10 g NaCl are dissolved in 800 mL water. Volume is adjusted to 1000 mL and autoclaved on the same day.
10. LB-agar: 5 g yeast extract, 10 g tryptone, 10 g NaCl, and 15 g agar are dissolved in 800 mL water. Volume is adjusted to 1000 mL and autoclaved on the same day.
11. 1000× Antibiotic stocks: Ampicillin (100 mg/mL); Kanamycin (50 mg/mL); Tetracycline (10 mg/mL in ethanol); Gentamycin (7 mg/mL), stored at −20 °C. All stocks prepared in water are filtered through a 0.22 μm syringe filter.

2.2 Bac-to-Bac Baculovirus Expression

1. Blue-gal (100 mg/mL in DMSO); IPTG (40 mg/mL) stored at −20 °C. All stocks prepared in water are filtered through a 0.22 and 0.45 μm syringe filter.
2. Bacmid transfection reagent.
3. Bacmid extract kit (plasmid mini kit (100)) (Qiagen).
4. Insect cell: Sf21 and High-5 (Invitrogen).
5. Insect cell media: Sf900II medium and High express five (Invitrogen); SIM HF (Sino Biological Inc).
6. 75 cm^2 flasks.

2.3 Test Expression

1. Lysis and wash buffer (I): 25 mM Tris–HCl, pH 7.5, 150 mM NaCl, 20 mM imidazole.
2. Elution buffer: 25 mM Tris–HCl, pH 7.5, 150 mM NaCl, 250 mM imidazole.
3. Ni-IDA Metal Chelate Resin (Qiagen).

2.4 Large-Scale Expression and Purification

1. Lysis and wash buffer(II): 25 mM Tris–HCl, pH 7.5, 1.5 M NaCl, 20 mM imidazole.
2. PreScission protease (PPase) was prepared in-house.
3. Econo-Columns.
4. Amicon Ultra protein concentrators (Millipore).

5. Size Exclusion Chromatography (SEC) buffer: 10 mM HEPES, pH 7.0, 100 mM NaCl. Filtered through a 0.2 μm membrane filter and stored at 4 °C.
6. Size-exclusion chromatography (Superdex-200) (GE healthcare).

2.5 ATPase Assay

1. [γ^{-32}P]ATP.
2. ATP is dissolved by water to 1 mM.
3. 5 × ATPase reaction buffer: 20 mM $MgCl_2$, 500 mM Tris–HCl, pH 8.0.
4. The thin-layer chromatography cellulose (TLC) plates.
5. ATPase reaction quenching buffer: 0.5 M EDTA.
6. TCL plates running buffer: 0.8 M acetic acid and 0.8 M LiCl.
7. Typhoon Trio Variable Mode Imager.

2.6 Helicase Assay

1. ATP, GTP, CTP, and TTP.
2. Trap RNA (5′-CGAAGCUGCUAACAUCAG-3′), top strand RNA (5′-UUUUUUUUUUCUGAUGUUAGCAGCUUC G-3′), and bottom stand RNA with a 5′-HEX tag (5′-HEX-C GAAGCUGCUAACAUCAG-3′).
3. The partial duplex RNA substrate with 5′ overhang is prepared by mixing the top strand RNA and the 5′-HEX tagged bottom strand.
4. 10× Helicase reaction buffer: 500 mM Hepes, pH 7.5, 50 mM $MgCl_2$, 20 mM DTT, and 0.1% BSA.
5. 5 × loading buffer: 100 mM Tris–HCl, pH 7.5, 50% glycerol, and 1% SDS.
6. 10× TBE buffer:108 g Tris, 7.44 g $Na_2EDTA.2H_2O$, 55 g boric acid, dissolved by water and adjusted the volume to 1 L.
7. 6% native PAGE gel: 1 mL 10 × TBE, 5.2 mL 30% polyacrylamide, 0.39 mL glycerol (50%), 13.25 mL H_2O, 0.15 mL ammonium persulfate (10%), and 0.01 mL TEMED.

2.7 Crystallization

1. Crystallization screen kits (Hampton and Qiagen).
2. 5′-triphosphate DNA (ppp-15T) are synthesized and purified according to previously published procedures [16, 17].
3. 24-Well vapor diffusion crystallization plates.
4. Crystallization conditions screen kits are supplied by Hampton research.

2.8 Structure Determination

The software that are used in structure determination include XDS, Coot, SHARP/autoSHARP, PHENIX, and Pymol.

3 Methods

All procedures should be carried out at room temperature unless otherwise specified.

3.1 Transfer Plasmid Construction and Transposition in E. coli DH10 Bac

1. Amplify MERS-nsp13 full-length by PCR method with BamHI and XhoI restriction sites at 5′ and 3′ termini, respectively.
2. The amplified MERS-nsp13 gene should be digested by BamHI/XhoI at 37 °C for 1 h. The pFast-bac-6×Histidine-SUMO plasmid should also be digested by BamHI/XhoI at the same time.
3. Digested nsp13 DNA should be ligated with pFast-bac-6×Histidine-SUMO vector using the rapid DNA ligation kit. The ligation system: 100 ng DNA, 35 ng vector, 10 × reaction buffer and T4 ligase, mixed and incubated at room temperate for 2 h.
4. Add 10 μL mixture of ligation product to 100 μL *E. coli* BL21 competent cells in a 1.5 mL Eppendorf tube and incubate on ice for 30 min. Heat shock the cells for 90 s in a 42 °C water bath and return briefly to ice. After 2 min, add 300 μL LB medium and incubate in a shaker at 37 °C and shake at 200 rpm for 1 h.
5. Spread the culture onto the LB plate containing Ampicillin (100 μg/mL), and incubate at 37 °C for 14–16 h.
6. Pick single colonies and inoculate to 500 μL LB medium. Incubate cultures at 37 °C and shake at 200 rpm for 4 h. Send the cultures for sequencing.
7. Collect the positive colonies and extract the recombinant plasmids according to the sequencing result.
8. Add 20 ng recombinant plasmid to 50 μL *E. coli* DH10 Bac competent cells in a 1.5 mL Eppendorf tube and incubate on ice for 30 min. Heat shock the cells for 90 s in a 42 °C water bath and return briefly to ice. 2 min later, add 500 μL LB medium and incubate in a shaker at 37 °C and shake at 200 rpm for 5 h.
9. Prepare bacmid selection plates containing approximately 10 mL of LB-agar, supplemented with 50 μg/mL kanamycin, 7 μg/mL gentamycin, 10 μg/mL tetracycline, 40 μg/mL IPTG, and 100 μg/mL Blue-gal, and once set allow to dry, inverted at room temperature.
10. Spread 50 μL culture onto the bacmid selection plate and incubate at 37 °C for up to 60 h.

3.2 Production of MERS-CoV nsp13 bacmid

1. Pick single white colony from the bacmid selection plate (white colonies contain the recombinant bacmid DNA and the blue ones do not). Inoculate it to 15 mL LB (containing 50 μg/mL kanamycin, 7 μg/mL gentamycin, 10 μg/mL tetracycline) and incubate at 37 °C with shaking at 220 rpm for up to 5 h.
2. Centrifuge the culture at *3000* × *g* for 20 min. Remove the supernatant carefully. Add 1.2 mL solution P1 from the bacmid extract kit (plasmid mini kit (100) from Qiagen) and resuspend the precipitate.
3. Add 1.2 mL solution P2, mix thoroughly by softly inverting 6–8 times, and incubate at room temperature (about 25 °C) for 5 min.
4. Add 1.2 mL solution P3, mix thoroughly by softly inverting 6–8 times. Incubate it on ice for 5 min.
5. Centrifuge at 15,000 × *g* for 10 min at 4 °C. Apply the supernatant to the QIAGEN-tip and allow it to enter the resin by gravity flow.
6. Wash the QIAGEN-tip with 2 × 2 mL Buffer QC. Elute DNA with 0.8 mL Buffer QF into a clean 1.5 mL Eppendorf tube.
7. Precipitate DNA with 0.56 mL isopropanol and wash the pellet with 1 mL 70% ethanol. Dry the pellet and resuspend in 20 μL sterilized ddH_2O.

3.3 Production of MERS-CoV nsp13 Recombinant Virus

1. Seed 0.8 × 10^6 Sf21 cells in duplicates in 6-well plates and incubate for 15–30 min at 28 °C (*see* **Notes 1** and **2**).
2. Prepare transfection reagent solution of 100 μL SF 900 II media (nonantibiotic) with 6 μL transfection reagents. Prepare bacmid solution of 100 μL SF 900 II media (nonantibiotic) with 4 μL dissolved bacmid. Mix the two solutions and rest at 28 °C for 30 min.
3. Add 800 μL SF 900 II media (nonantibiotic) to the mixture, and transfer them to the 6-well plates in **step1** (after removing supernatant and washed by nonantibiotic media twice). Prior to addition of SF90II media to the 6-well plates make sure to remove the supernatant and wash twice with nonantibiotic media. Incubate the 6-well plate at 28 °C for 5 h.
4. Remove the supernatant of the plates and add 2.5 mL fresh media with 10 μg/mL gentamycin. Incubate at 28 °C for 72 h. Observe the cells under the microscope. Collect the supernatant if the cells present noticeable infected symptoms (swelling, splitting, and stop growing), and centrifuge at 453 × *g* for 5 min. Collect the supernatant. This will be P1 virus.
5. Prepare 75 cm^2 flasks containing 15 mL Sf21 cell suspension at a density of 0.4–0.6 × 10^6 cells/mL. Make sure the cells are distributed evenly in the flasks, incubate at 28 °C for 20 min.

6. Add 0.4 mL P1 virus to the 15 mL Sf21 cells culture in 75 cm^2 flask. Incubate at 28 °C for 48–60 h. Observe the cells under the microscope and collect the supernatant from the flasks, centrifuge at 453 × *g for* 5 min. Collect the supernatant. This will be P2 virus.
7. Prepare 75 cm^2 flasks containing 15 mL Sf21 cells suspension at a density of 0.6–1.0 × 10^6 cells/mL. Make sure the cells are distributed evenly in the flask, incubate at 28 °C for 20 min.
8. Infect the 15 mL Sf21 cell culture in 75 cm^2 flask with 0.4 mL P1 virus. Incubate at 28 °C for 48–60 h. Observe the cells under the microscope and collect the supernatant from the flasks, centrifuge at 453 × *g* for 5 min. Collect the supernatant (P3 virus).

3.4 Test Expression of MERS-CoV nsp13

1. Prepare 50 mL high-5 cells in express-5 medium at a density of 0.38 × 10^6 cells/mL, and culture in a 300 mL cell conical flask. Incubate the culture at 28 °C with shaking at 120 rpm for 48 h, the density of cells will grow to 1.5–2.5 × 10^6 cells/mL (*see* **Note 2**).
2. Add 1.5 mL MERS-CoV nsp13 P2 or P3 virus into the culture, and incubate at 22 °C with shaking at 120 rpm for 44–60 h.
3. Centrifuge the culture at 3000 × *g* for 30 min. Collect the cells pellet.
4. Quickly freeze the cells pellet by liquid nitrogen. Resuspend the pellet by 5 mL lysis and wash buffer (I) and incubate on ice for up to 10 min.
5. Centrifuge at 15000 × *g* for 20 min at 4 °C. Apply the supernatant to the mini-affinity column with 300 μL Ni-IDA Metal Chelate Resin and allow it to enter the resin by gravity flow at 4 °C.
6. Wash the resin with 1 mL lysis and wash buffer (I) three times.
7. Load 600 μL elution buffer to the resin and allow it to enter the resin by gravity flow. Collect the eluted sample into a 1.5 mL Eppendorf tube.
8. Pick 8 μL eluted sample and mix with 2 × loading buffer.
9. Load SDS-PAGE gel and run at 200 V for 60 min. The result of expression is visualized by Coomassie brilliant blue stain (Fig. 1).

3.5 Large-Scale Expression and Purification of MERS-CoV nsp13

1. Prepare 1 to 1.5 L high-5 cells in HF medium at a density of 0.38 × 10^6 cells/mL, and culture in 3 L cell conical flask (keep the volume of culture to 500–750 mL in a 3 L flask). Incubate the culture at 28 °C with shaking at 120 rpm for 40–48 h, then to the density of cells will grow to 1.5–2.5 × 10^6 cells/mL (*see* **Note 3**).

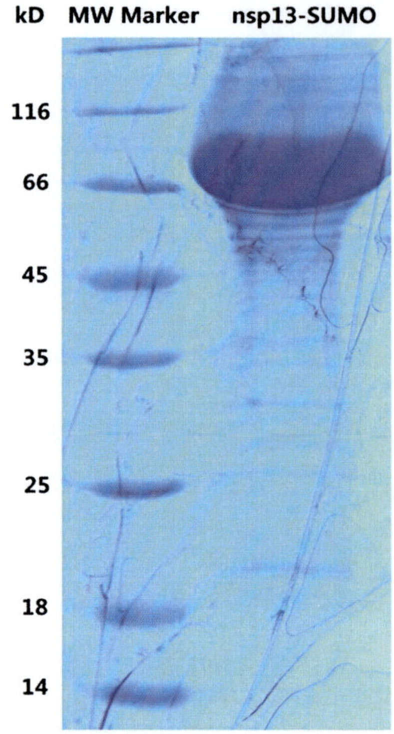

Fig. 1 Test expression of MERS-CoV nsp13. The eluted MERS-CoV nsp13 protein possesses an 6-Histidine and SUMO tag. 8 μL sample was loaded on the SDS-PAGE gel and the result was visualized by Coomassie brilliant blue stain

2. Add 30–45 mL MERS-CoV nsp13 P3 virus to the culture (30 mL virus/L culture). Incubate the culture at 22 °C with shaking at 120 rpm for 40–48 h.
3. Centrifuge the culture at 3000 × *g* for 30 min. Collect the cell pellet.
4. Resuspend the pellet by 100–150 mL lysis and wash buffer (II) (100 mL buffer/1 L culture's pellet). Add 600–900 μL 0.1 M PMSF to the cell suspension (600 μL PMSF/100 mL cell suspension).
5. Place the cell suspension on ice-water mixture. Set the amplitude to 30% on a 750 W cell sonicator and sonicate with bursts of 3 s on, 5 s off.
6. Transfer the lysates to centrifuge tubes, balance the tubes pairwise and centrifuge at 15000 × *g* for 1 h at 4 °C.
7. Transfer the supernatant to new centrifuge tubes and re-centrifuge at 15,000 × *g* for 1 h at 4 °C.
8. Transfer the clear supernatant into clean tubes taking care to avoid transferring any pelleted material.
9. Filter the supernatant by 0.45 μm syringe filter. This clarified supernatant represents the soluble fraction.

10. Prepare the Ni-NTA resin, and add the resin into 2–3 empty Econo-Columns (5 mL 50% resin per column), wash and balance the resin with 10 mL lysis and wash buffer(II) twice.
11. Place the columns at 4°C. Apply the clarified cell lysates supernatant to the balanced Ni-NTA resin, and flow through the column by gravity.
12. Wash the resin in the column with 10 mL lysis and wash buffer (II) 3 times.
13. Resuspend the resin by 3.5 mL L lysis and wash buffer(II), and add 100 μL PPase. Incubate the resin at 4 °C for 10–12 h.
14. Apply the buffer to the column and let it flow under gravity. Collect the flow through in a 50 mL tube.
15. Add another 25 mL lysis and wash buffer(II) to the resin and flow through the column. Also collect the flow through in the previous 50 mL tube.
16. To remove the PPase, add the flow through to another column which contains the NS4B resin. Collect the flow through from the NS4B resin column.
17. Apply the flow through to an Amicon Ultra protein concentrator (30 kDa filter, 50 mL), centrifuge at 2465 × *g* at 4 °C until the sample volume is concentrated to 1 mL.
18. Transfer the concentrated sample to a 1.5 mL tube and centrifuge at 17,949 × *g* for 3 min to remove the aggregates and particulates.
19. Load the sample onto the superdex 200 column in the size exclude chromatography (SEC) buffer using an ÄKTA-purify chromatography at 4 °C.
20. Analyze 8 μL of each peak fractions by SDS-PAGE (Fig. 2).
21. Collect the fractions that contain the single band of MERS-CoV nsp13, mix the fractions, and concentrate the mixture to a final density of 6–8 mg/mL.
22. 50 μL packaged the protein sample, quickly freeze them by liquid nitrogen and store them at −80 °C.

3.6 ATPase Assay of MERS-CoV nsp13

1. Dilute the purified MERS-CoV nsp13 to 0.5 μM by SEC buffer.
2. Add the following reagents in turn to prepare the reaction mixture: ddH_2O (36.5 μL), 5 × ATPase reaction buffer (10 μL), ATP (1 mM, 2.0 μL), and [γ-^{32}P]ATP (~1 nM, 1 μL) [18].
3. Add diluted nsp13 protein (0.5 μM, 2 μL) to the reaction mixture, incubate at 30 °C and start timing.
4. At each indicated time point, add 2 μL quenching buffer (0.5 M EDTA) to the mixture to stop the reaction and place the mixture on ice.

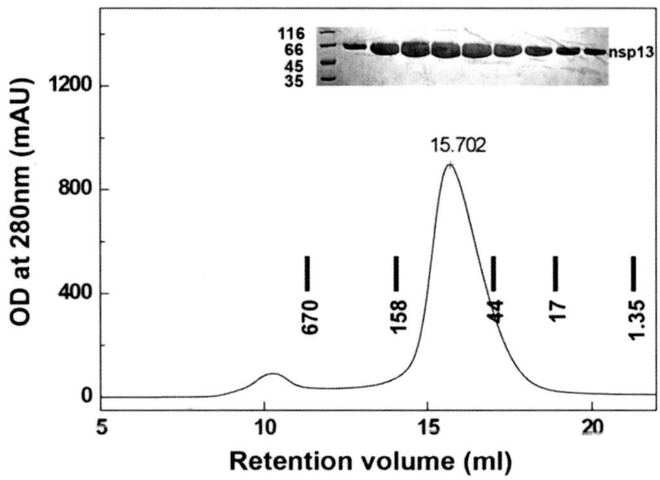

Fig. 2 Large-scale purification of MERS-CoV nsp13.MERS-CoV nsp13 eluted from Superdex 200 300/10 GL column precalibrated with gel filtration standards (thyroglobulin 670 kDa, γ-globulin 158 kDa, ovalbumin 44 kDa and myoglobin 17 kDa and vitamin 1.35 kDa (1350 Da)). Upper insert, SDS-PAGE analysis of the purified protein

5. Spot 1 μL sample from the mixture on the thin-layer chromatography cellulose TLC plates and resolve with running buffer for 20 min.
6. Dry the plates and press the plate onto phosphor screen for 2 h. Analyze the result by storage phosphor screen and Typhoon Trio Variable Mode Imager (Fig. 3).

3.7 Helicase Assay of MERS-CoV nsp13

1. Dilute the purified MERS-CoV nsp13 to 1 μM by SEC buffer (*see* **Note 4**).
2. Add the following reagents in turn: 10 × helicase reaction buffer (1 μL), H_2O (4 μL), Trap RNA (3 μM, 1 μL), partial duplex RNA substrate (0.5 μM, 1 μL), and diluted nsp13 protein (0.5 μM, 2 μL), ATP (10 mM, 1 μL). The final volume of each reaction mixture is 10 μL [19].
3. Incubate the mixtures at 30 °C for 30 min.
4. Add 2.5 μL 5 × loading buffer to the mixture to stop the reaction.
5. Take 4 μL sample from each reaction mixtures and load the samples onto 10% native PAGE gel.
6. Run the native PAGE gel at 100 V for 40 min on ice.
7. Scan the gel (Fig. 4).

3.8 Crystallization of MERS-CoV nsp13

Crystals of the unliganded MERS-CoV nsp13 diffracted the X-rays poorly, >3.6 Å. The addition of 5′-triphosphate-15 dT DNA (ppp-15T) greatly improves the resolution.

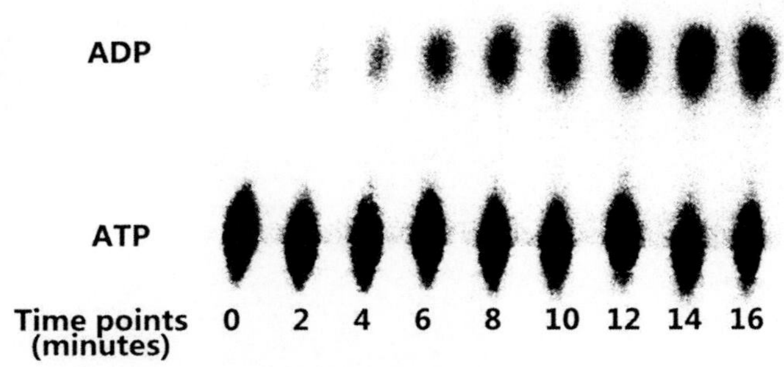

Fig. 3 ATPase activity of MERS-CoV nsp13. ATPase activity was measured by incubation at 30 °C for 0–16 min

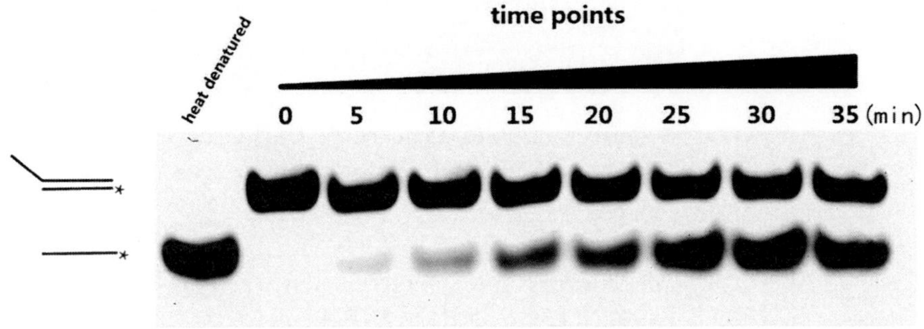

Fig. 4 Helicase activity of MERS-CoV nsp13. Activity was determined with the indicated RNA substrate (the asterisk marks the position of the HEX label). Samples were incubated at 30 °C for 0–35 min

1. Mix the purified MERS-CoV nsp13 with 5′-triphosphate-15T DNA (ppp-15T) with 1:1.5 molar ratio and incubate at 4 °C overnight.
2. Mix 1 μL sample with 1 μL reservoir buffer from the crystallization conditions screen kits, and incubate at 18 °C using the hanging-drop vapor-diffusion system.
3. Crystallize MERS-CoV nsp13 by mixing with the equal volume of reservoir buffer containing 0.1 M Tris–HCl (pH 8.5), 1 M $(NH4)_2SO_4$, and 15% glycerol. Crystals grow to their maximum in a week (Fig. 5).

3.9 Determination of MERS-CoV nsp13 Crystal Structure

1. Highly redundant multi-wavelength anomalous diffraction data should be collected using the X-ray with wavelengths close to the absorption edge of zinc. High energy remote wavelength should be 1.2810 Å, peak wavelength: 1.2827 Å (two datasets were collected to improve the redundancy), and inflection wavelength 1.2831 Å.
2. Data processing and reducing by XDS Package and Truncate software from CCP4. The crystals belong to the space group $P6_122$, and contained two copies of nsp13 per asymmetric unit.

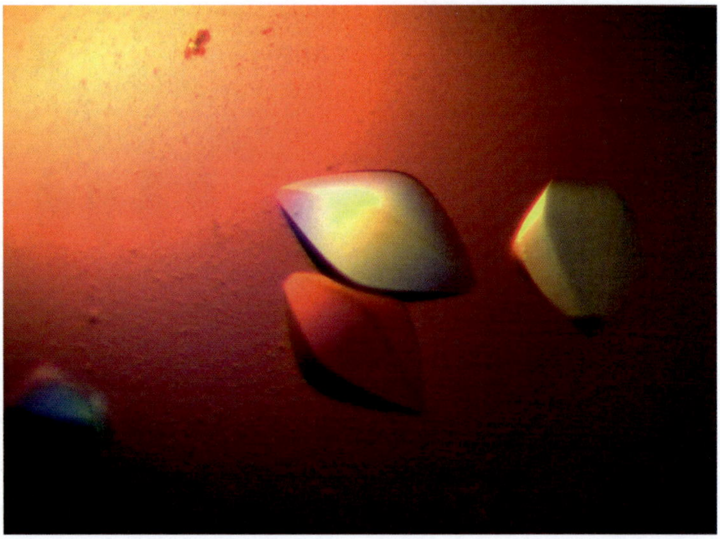

Fig. 5 Crystals of MERS-CoV nsp13. Nsp13 crystallized by incubation with 5′-triphosphate-15T DNA at 18°C using the hanging-drop vapor-diffusion system. The crystals are grown to the biggest in a week

3. An interpretable electron density map should be calculated using SHARP/autoSHARP [20].
4. Manually build the initial model of MERS-CoV nsp13 by Coot [21].
5. Collect native data with highest resolution (3.0 Å) using the X-rays with the wavelength of 0.978 Å.
6. Higher resolution structure should be solved by molecular replacement using the initial nsp13 structure as the searching model.
7. Manual model building with the improved electron density map. While most part of nsp13 can be located, the electron density of 1B subdomain is very weak, reflecting that this part is highly flexible.
8. Structure refinement to resolution limit of 3.0 Å using software PHENIX [22].

In the final model (Fig. 6), 145-230aa (the entire 1B domain) of molecule A are disordered, probably due to mobility of 1B and the lack of crystal contacts, whereas in molecule B, 591 out of 598 amino acids were located in the electron density maps (Fig. 7). Data collection and refinement statistics are summarized in Table 1.

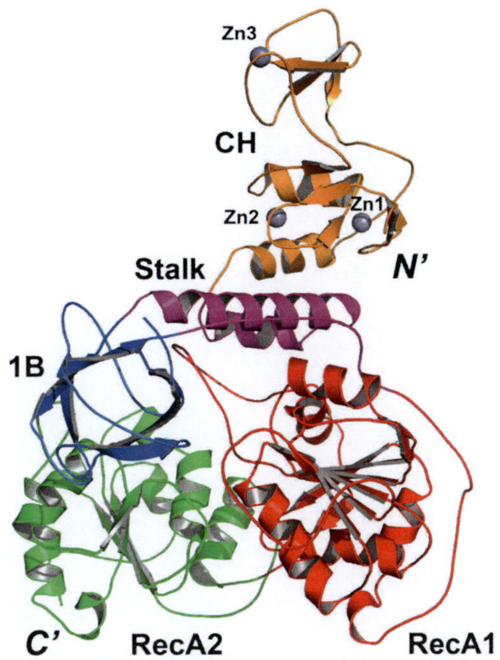

Fig. 6 Final model of MERS-CoV nsp13 structure. Model of MERS-CoV nsp13 containing CH (orange), Stalk (magenta), 1B (blue), RecA1 (red), and RecA2 (green) domains

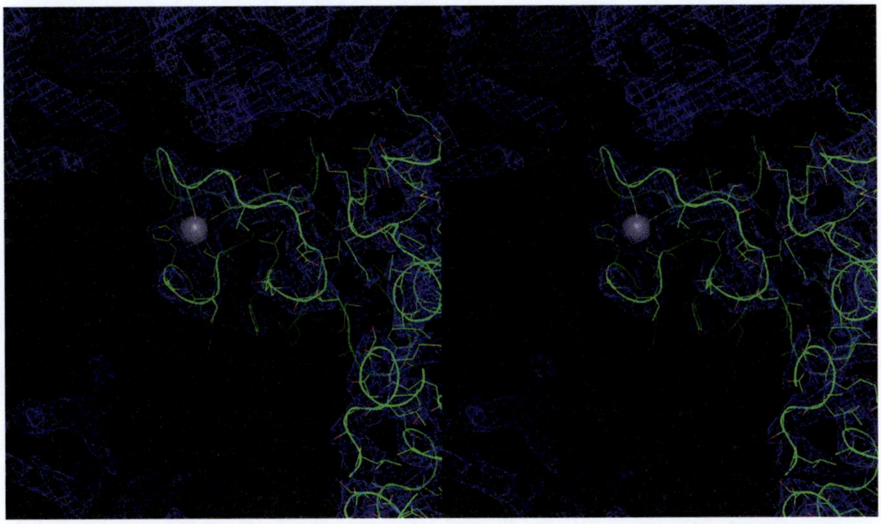

Fig. 7 Portion of the electron density map of MERS nsp13 crystal structure. A wall-eye stereo image of a portion of electron density map (zn3 binding site). 2Fo-Fc map is shown with blue mesh. The final model of MERS nsp13 (green) is superimposed. The zinc is shown with a gray sphere

Table 1
Data collection and refinement statistics

	MERS-CoV nsp13			MERS-CoV nsp13 (PDB ID: 5WWP)
Data collection	MAD phasing			Native data
Space group	$P6_122$			$P6_122$
Cell dimensions				
a, b, c (Å)	186.19, 186.19, 185.44			185.68, 185.68, 185.09
α, β, γ (°)	90.00, 90.00, 120.00			90.00, 90.00, 120.00
X-ray source	SLS X06DA			SSRF BL19U1
Wavelength (Å)	1.2827 (peak)	1.2831 (Infl)	1.2810 (Hrem)	0.9784 (native)
Data range (Å)	49.06–3.12	49.09–3.12	49.05–3.12	48.95–3.00
Reflections unique	63,859	64,027	63,771	71,207
R_{sym}[a] (last shell)	0.29 (1.89)	0.29 (2.01)	0.30 (1.98)	0.08 (0.75)
CC(1/2)	99.7 (61.6)	99.7 (59.6)	99.3 (47.7)	99.8 (57.1)
$I/\sigma I$	13.80 (1.76)	13.64 (1.67)	10.66 (1.31)	12.27 (1.57)
Completeness (%) (last shell)	99.9 (99.7)	99.9 (99.8)	99.9 (99.7)	99.4 (98.4)
Redundancy (last shell)	21.15 (19.79)	21.14 (19.77)	12.91 (12.08)	3.38 (3.36)
Refinement				
Resolution range (Å)				48.95–3.00
% reflections in cross-validation				4.81
R_{work}[b]/R_{free}[c] (last shell)				0.23, 0.28 (0.38, 0.41)

Atoms	
All atoms	8571
Protein	8540
Zinc	6
Solvent	25
B-factors average (Å^2)	68.1
Protein (Å^2)	68.1
Ligands (Å^2)	68.5
Solvent (Å^2)	90.2
r.m.s.d	
Bond lengths (Å)	0.015
Bond angles (°)	0.950
Validation	
MolProbity score	2.75, 88th percentile [d]
Clashscore, all atoms	15.39, 97th percentile [d]
% residues in favored regions, allowed regions, outliers in Ramachandran plot	91.8, 7.5, 0.7

[a] $R_{sym} = \Sigma_{hkl}\Sigma_j |I_{hkl,j} - I_{hkl}|/\Sigma_{hkl}\Sigma_j I_{hkl,j}$, where I_{hkl} is the average of symmetry-related observations of a unique reflection

[b] $R_{work} = \Sigma_{hkl}||F_{obs}(hkl)| - |F_{calc}(hkl)||/\Sigma_{hkl}|F_{obs}(hkl)|$

[c] R_{free} = the cross-validation R factor for 5% of reflections against which the model was not refined

[d] 100th percentile is the best among structures of comparable resolution; 0th percentile is the worst. For clash score the comparative set of structures was selected in 2004, for MolProbity score in 2006

4 Notes

1. When we prepare P1 virus in six-well plates, the medium in the wells always evaporated. Sealing the gap of the plate by medical tape can reduce the evaporation of medium (don't seal the gap completely, leave a small gap to keep the ventilation). Having a water trough in incubator also can reduce the evaporation of the medium.
2. The culture of insect cells sometimes was harassed by the contamination of bacteria or other microbes. To avoid the contamination, we treat the conical flasks not only by conventional autoclave sterilization, but also leave the 3 L conical flask (sealed by tinfoil) in the oven at 200 °C for 3–5 h before using.
3. To remove nucleic acids bound to nsp13, we used the lysis buffer containing high concentrate salt; this is a key step and improves the crystallization of nsp13 [10]. In practice, when sonicated in the buffer containing high concentrate salt, we found that the SUMO-tagged recombinant proteins lead the supernatant of the high-5 cell lysate to be turbid, which finally blocks the affinity columns. We have tried four concentrations of NaCl in lysis buffer, including 300 mM, 500 mM, 1 M, and 1.5 M. The first three concentrations of NaCl render the supernatant to be unable to use, we can't improve it by high-speed centrifugation (47,850 × *g*), and it also can't be filtered by 0.45 μm syringe filter. The last concentration, 1.5 M NaCl in lysis buffer, could generate a bit better supernatant of cell lysates than other three concentrations of salt. We centrifuge the supernatant twice, then can filter it by 0.45 μm syringe filters (100 mL supernatant consumed about 8–10 filters). This clarified supernatant can flow through the affinity columns well.
4. The results of helicase assay always face the contamination of background fluorescence. Keep the gel from contacting any items containing fluorescence in the lab, including fluorescent dyes, some plastic boxes, hand towel, and so on.

References

1. Stadler K, Masignani V, Eickmann M, Becker S, Abrignani S et al (2003) SARS--beginning to understand a new virus. Nat Rev Microbiol 1:209–218
2. Zaki AM, van Boheemen S, Bestebroer TM, Osterhaus AD, Fouchier RA (2012) Isolation of a novel coronavirus from a man with pneumonia in Saudi Arabia. N Engl J Med 367:1814–1820
3. Kupferschmidt K (2015) INFECTIOUS DISEASES. Amid panic, a chance to learn about MERS. Science 348:1183–1184
4. Gorbalenya AE, Enjuanes L, Ziebuhr J, Snijder EJ (2006) Nidovirales: evolving the largest RNA virus genome. Virus Res 117:17–37
5. Lauber C, Goeman JJ, Parquet Mdel C, Nga PT, Snijder EJ et al (2013) The footprint of genome architecture in the largest genome

expansion in RNA viruses. PLoS Pathog 9: e1003500

6. Thiel V, Ivanov KA, Putics A, Hertzig T, Schelle B et al (2003) Mechanisms and enzymes involved in SARS coronavirus genome expression. J Gen Virol 84:2305–2315
7. Subissi L, Imbert I, Ferron F, Collet A, Coutard B et al (2014) SARS-CoV ORF1b-encoded nonstructural proteins 12-16: replicative enzymes as antiviral targets. Antivir Res 101:122–130
8. Subissi L, Posthuma CC, Collet A, Zevenhoven-Dobbe JC, Gorbalenya AE et al (2014) One severe acute respiratory syndrome coronavirus protein complex integrates processive RNA polymerase and exonuclease activities. Proc Natl Acad Sci U S A 111: E3900–E3909
9. Prentice E, McAuliffe J, Lu X, Subbarao K, Denison MR (2004) Identification and characterization of severe acute respiratory syndrome coronavirus replicase proteins. J Virol 78:9977–9986
10. Hao W, Wojdyla JA, Zhao R, Han R, Das R et al (2017) Crystal structure of Middle East respiratory syndrome coronavirus helicase. PLoS Pathog 13:e1006474
11. Gorbalenya AE, Koonin EV (1989) Viral proteins containing the purine NTP-binding sequence pattern. Nucleic Acids Res 17:8413–8440
12. Deng Z, Lehmann KC, Li X, Feng C, Wang G et al (2014) Structural basis for the regulatory function of a complex zinc-binding domain in a replicative arterivirus helicase resembling a nonsense-mediated mRNA decay helicase. Nucleic Acids Res 42:3464–3477
13. Ivanov KA, Thiel V, Dobbe JC, van der Meer Y, Snijder EJ et al (2004) Multiple enzymatic activities associated with severe acute respiratory syndrome coronavirus helicase. J Virol 78:5619–5632
14. Adedeji AO, Lazarus H (2016) Biochemical characterization of Middle East respiratory syndrome coronavirus helicase. mSphere 1. https://doi.org/10.1128/mSphere.00235-16
15. Hu Z, Yan C, Liu P, Huang Z, Ma R et al (2013) Crystal structure of NLRC4 reveals its autoinhibition mechanism. Science 341:172–175
16. Zlatev I, Lackey JG, Zhang L, Dell A, McRae K et al (2013) Automated parallel synthesis of 5′-triphosphate oligonucleotides and preparation of chemically modified 5′-triphosphate small interfering RNA. Bioorg Med Chem 21:722–732
17. Zlatev I, Manoharan M, Vasseur JJ, Morvan F (2012) Solid-phase chemical synthesis of 5-′-triphosphate DNA, RNA, and chemically modified oligonucleotides. Curr Protoc Nucleic Acid Chem. Chapter 1: Unit1 28
18. Cui S, Eisenacher K, Kirchhofer A, Brzozka K, Lammens A et al (2008) The C-terminal regulatory domain is the RNA 5′-triphosphate sensor of RIG-I. Mol Cell 29:169–179
19. Lee NR, Kwon HM, Park K, Oh S, Jeong YJ et al (2010) Cooperative translocation enhances the unwinding of duplex DNA by SARS coronavirus helicase nsP13. Nucleic Acids Res 38:7626–7636
20. Vonrhein C, Blanc E, Roversi P, Bricogne G (2007) Automated structure solution with autoSHARP. Methods Mol Biol 364:215–230
21. Emsley P, Lohkamp B, Scott WG, Cowtan K (2010) Features and development of coot. Acta Crystallogr D Biol Crystallogr 66:486–501
22. Adams PD, Afonine PV, Bunkoczi G, Chen VB, Davis IW et al (2010) PHENIX: a comprehensive python-based system for macromolecular structure solution. Acta Crystallogr D Biol Crystallogr 66:213–221

PART III

Quantitation of Virus and Anti-Viral Factors

Chapter 7

Antigen Capture Enzyme-Linked Immunosorbent Assay for Detecting Middle East Respiratory Syndrome Coronavirus in Humans

Joshua Fung, Susanna K. P. Lau, and Patrick C. Y. Woo

Abstract

The Middle East respiratory syndrome (MERS) is the second novel zoonotic disease infecting humans caused by coronavirus (CoV) in this century. To date, more than 2200 laboratory-confirmed human cases have been identified in 27 countries, and more than 800 MERS-CoV associated deaths have been reported since its outbreak in 2012. Rapid laboratory diagnosis of MERS-CoV is the key to successful containment and prevention of the spread of infection. Though the gold standard for diagnosing MERS-CoV infection in humans is still nucleic acid amplification test (NAAT) of the up-E region, an antigen capture enzyme-linked immunosorbent assay (ELISA) could also be of use for early diagnosis in less developed locations. In the present method, a step-by-step guide to perform a MERS-CoV nucleocapsid protein (NP) capture ELISA using two NP-specific monoclonal antibodies is provided for readers to develop their in-house workflow or diagnostic kit for clinical use and for mass-screening project of animals (e.g., dromedaries and bats) to better understand the spread and evolution of the virus.

Key words MERS-CoV, Nucleocapsid protein, Molecular detection, Clinical diagnosis, Antigen capture ELISA, Immunoassay

1 Introduction

The Middle East respiratory syndrome (MERS) is the second novel zoonotic disease infecting humans caused by coronavirus (CoV) in this century. To date, more than 2200 laboratory-confirmed human cases have been identified in 27 countries, and more than 800 MERS-CoV associated deaths have been reported since its outbreak in 2012 [1]. Rapid laboratory diagnosis of MERS-CoV is the key to successful containment and prevention of the spread. Nucleic acid amplification test (NAAT, e.g., real-time reverse transcription quantitative polymerase chain reaction [real-time RT-qPCR]), virus isolation, transmission electron microscopy, immunohistochemistry, and serological methods (e.g., antigen capture enzyme-linked immunosorbent assay [ELISA] and

Rahul Vijay (ed.), *MERS Coronavirus: Methods and Protocols*, Methods in Molecular Biology, vol. 2099,
https://doi.org/10.1007/978-1-0716-0211-9_7,

immunofluorescence assay [IFA]) have been developed and used for MERS-CoV diagnosis [2–7]. While the "gold standard" for MERS-CoV diagnosis is NAAT of the upper region of the envelope gene (up-E) or the nucleocapsid (N) gene as suggested by the World Health Organization (WHO), antigen capture ELISA assay for MERS-CoV can also be informative when NAAT is not available or when the serological assay is used to confirm the findings and aid treatment decision [2, 3].

Further to diagnosing possible human infection of MERS-CoV, this method is also useful for screening the virus in the wildlife or agricultural applications. Government agencies and research groups may find serological tests like antigen capture ELISA to be more economical than NAATs for routine screening of MERS-CoV in farm-held or city-dwelling animals. The antigen capture ELISA described in this method offers four significant advantages over traditional NAATs.

Firstly, serological screening requires less space in facilities and can be performed in point-of-care locations to minimize sample transporting and reduce turnover time. To avoid cross-contamination from amplicons in NAATs, the workflow usually requires four separate physical locations: (1) sample preparation (lysis, extraction of nucleic acids, and reverse transcription), (2) NAAT master mix preparation, (3) template addition, and (4) amplification and analysis. Though technologies like real-time RT-qPCR simplify the workflow, such requirements limit the assay to be performed in regional laboratories designed or designated for this application. Antigen capture ELISA, on the other hand, can be performed on open bench in a single location after virus inactivation, allowing it to be performed in even the most minimally designed facility.

Secondly, antigen capture ELISA can be performed with simple equipment and can be established with limited initial investment. For performing NAATs at a modern standard, UV cabinets or workstations for master mix preparation and sample addition, thermal cyclers, agarose gel running, and visualization equipment are the least requirement. For more stringent testing and faster turnaround, it calls for a real-time PCR thermal cycler (e.g., Roche's LightCycler systems or Bio-Rad Touch detection systems) which requires a fair amount of initial investment and limits the assay from being performed in remote or less developed locations. In contrast, antigen capture ELISA and other serological methods can be performed with much simpler equipment. Multichannel pipettes, automatic plate washer, and plate reader are the only specialized tools needed for this application and can be purchased with ease if those are not already available.

Thirdly, much less training is required for technicians to handle serological testing than NAATs. Though NAATs and ELISA are some of the most basic assays performed in a medical laboratory and minimal training is needed for an experienced worker to perform such task, to allow quicker and broader surveillance of MERS-CoV in human and animal population, it would be beneficial to set up more surveillance facilities in the less developed parts of the world. The time and resources needed to train a novice laboratory worker to perform ELISA are much less, as only dilution and pipetting skills are required.

Fourthly, common nucleic staining chemicals used in NAATs for amplicon visualization are a possible mutagen and post potential health risk to workers and the surrounding environment; while chemicals and solutions used in ELISA are relatively safer. To visualize the amplicons after agarose gel electrophoresis or during the qPCR thermal cycles, dyes like ethidium bromide (EtBr), SYBR Green, or Gel Red are used; while EtBr is a known mutagen, others are a relatively new addition to the market and extensive safety data is not widely available [8]. In comparison, the chemicals and solutions used in ELISA are commonly found in clinical and research laboratories and are generally safe when used properly.

Finally, and most importantly, antigen capture ELISA can offer high sensitivity and specificity for MERS-CoV diagnosis in even early infection and animal samples. We have previously demonstrated that by using two MERS-CoV nucleocapsid protein (NP) specific monoclonal antibodies (MAbs) in performing capture ELISA, the test can accurately detect MERS-CoV virus down to 10 $TCID_{50}$/0.1 mL and has a specificity of 100% [3]. As the nasopharyngeal aspirate viral load from patients during acute infection are around 10^6 copies/mL and nasal samples in dromedaries are usually around 10^4–10^6 copies/mL, this test offers sufficient sensitivity for MER-CoV diagnosis and screening [9–11].

Other forms of MERS-CoV serological diagnostic test have also been developed based on different principles and are designed to fulfill different purposes, one should also review those options and evaluate their needs. To detect seroconversion from previous infection of MERS-CoV, the WHO suggests laboratories to perform IFA or ELISA together with neutralization assay, the result alone can be used to determine if it is a confirmed case, regardless of the results from NAAT assay [2]. For rapid on-site diagnosis of MERS-CoV, we have previously reported the adaptation of the antigen capture ELISA in the format of lateral flow immunoassay (LFIA). This assay can yield results in under half an hour, requires minimal equipment, training, and can be stored at room temperature, thus allowing it to be performed in the field [12]. This LFIA is also able to detect MERS-CoV-like viruses (e.g., *Tylonycteris* bat CoV HKU4 and *Pipistrellus* bat CoV HKU5) and is useful for the research to understand the evolutional history of MERS-CoV [13, 14].

In the current manuscript, the method for performing NP capture ELISA using two MERS-CoV-NP-specific monoclonal

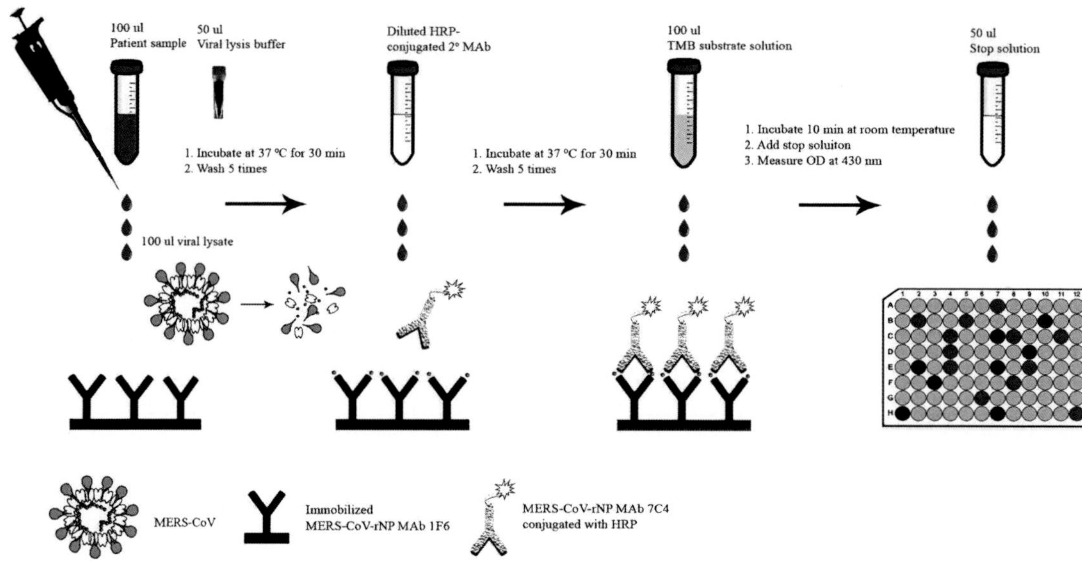

Fig. 1 Schematic diagram showing the general workflow of the MERS-CoV NP antigen capture ELISA

antibodies (MAbs) will be introduced. The general workflow of the assay is summarized in a figure for quick referencing [3, 15] (Fig. 1).

2 Materials

2.1 Reagents and Equipment

1. 1.5 mL conical screw cap tubes.
2. 10, 200, 300 (optional), and 1000 μL filtered pipette tips.
3. Single-channel (covering 10–1000 μL) and 8-channel (200 μL or 300 μL; optional) pipettes.
4. 96-Well high binding microtiter plates or strips with holder for ELISA.
5. Adhesive sealing film for microtiter plates.
6. 50 mL solution reservoir for multichannel pipettes.
7. Automated microtiter plate washer-dispenser (able to handle 96-well plates and microwell strips; optional) (*see* **Note 1**).
8. Microtiter plate spectrophotometer able to read optical density (OD) at 450 nm.
9. Platform rocker.
10. Two purified MERS-CoV NP MAbs with nonoverlapping epitopes.
11. TMB solution.
12. MAb 7C4 conjugated with HRP.
13. 3,3′,5,5′-Tetramethylbenzidine (TMB) substrate solution.

2.2 Buffers

1. Phosphate-buffered saline (PBS): 144 mg potassium phosphate monobasic, 9000 mg sodium chloride, and 795 mg of sodium phosphate dibasic salts in 1 L of water.
2. Washing buffer: PBS containing 0.5% Tween 20.
3. Blocking buffer: PBS containing 2% sucrose, 0.2% casein-Na, and 2% gelatin.
4. Enzyme dilution buffer: PBS containing 0.5% Tween 20 and 20% fetal calf serum.
5. Sample dilution buffer: PBS containing 2% skim milk.
6. Stop solution: 0.2 M sulfuric acid.
7. Viral Lysis Buffer (*see* **Note 4**).

3 Methods

3.1 Designing the Assay

The antigen capture ELISA is also known as sandwich ELISA and makes use of a "capture" antibody and a "detection" antibody. The capture antibody is coated onto the wells of a microtiter plate before the assay. Then following sample processing, the lysate is incubated in the wells of the microtiter plate. If the sample contains peptides from MERS-CoV (specifically nucleocapsid protein), they will bind with the coated antibody and be "captured" onto the microtiter plate. Even minute amount of viral peptide can be retained in the well if the capture antibody has a high affinity to the peptide and was coated at high concentration. Unbonded proteins are then washed away before the addition of the second, "detection" antibody. The secondary MAb also recognizes the MERS-CoV NP, presumably binds to a distinct epitope, and is conjugated with horseradish peroxidase for detection. The combination of two MAbs in an ELISA assay offers increased sensitivity for MERS-CoV NP. On the other hand, this "sandwich" approach also allows improved specificity for the MERS-CoV nucleocapsid protein by combining the specificities of the two MAbs, allowing it to differentiate and identify MERS-CoV spiked sample from other samples from healthy and patients who contracted various respiratory tract infections, as previously demonstrated [3].

In this assay the nucleocapsid protein was selected as the target for generating antibodies to detect MERS-CoV. According to previous experience when working with SARS-CoV, we observed that the NP is a highly immunogenic and abundantly expressed structural protein, and a more preferable target than the spike (S) protein [16, 17]. Working with the hypothesis that the NP protein of MERS-CoV might also be a desirable target when developing an antigen capture ELISA for it, we have shown that the assay offers high specificity and sensitivity, as mentioned above. The steps related to the cloning and purification of $(His)_6$-tagged recombinant NP (rNP) of MERS-CoV for the generation of anti-MERS-CoV-rNP MAbs will not be described, as there are commercially available antibodies readily available for purchase. The horseradish peroxidase (HRP) system was used for the colorimetric visualization at the final stage of the assay. Commercial ELISA kits may utilize other detection methods; optimization may be needed. For readers who would like to generate their own HRP conjugated detection antibody, there are also kits available.

3.2 Preparing Solutions

When preparing solutions and buffers for the assay, investigators should be aware that "old" buffers may be more likely to be contaminated. The accuracy and reproducibility of the assay can be affected, as the peptides from fungus or other microorganisms may compete with the target antigen. Prepare fresh solutions periodically (~1 month); autoclave or filter sterilize the buffers if available. If contaminations are a common occurrence, the addition of 0.05% sodium azide (NaN_3) as a preservative is an option.

3.3 Coating Microtiter Plates with Antibody

1. Dilute the MERS-CoV-rNP MAb 1F6 in blocking buffer. (*see* **Note 2**).
2. Coat the microtiter plates by adding 100 μL of the solution prepared per well.
3. Cover the plate with an adhesive plastic cover and incubate at 37 °C overnight (*see* **Note 3**).
4. Discard the adhesive plastic cover and remove the solution.
5. Wash the plate with 300 μL of washing buffer per well for five times using an automatic microplate washer.
6. Dry the plate by patting the plate on a paper towel.
7. Allow the plate to air-dry. Proceed to the next step or cover the plate with adhesive plastic cover and store at 4 °C until use.

3.4 Sample Processing

All processes with potentially infectious MERS-CoV materials should be handled according to institutional, local, and international regulations, guidelines, and standard operating procedures (SOP) to avoid spreading and contamination of the facility. All work with infectious MERS-CoV was performed inside a biosafety level-2 cabinet with SOP in approved biosafety level-3 facilities during development and evaluation of the assay [3, 18, 19].

1. Aliquot 50 μL of viral lysis buffer to new 1.5 mL conical screw cap tubes according to the number of samples and controls (*see* **Note 4**).
2. Pipette 100 μL of specimen from the sample collection tube to the 1.5 mL conical screw cap tubes with viral lysis buffer, mix well. Allow sufficient time for inactivation.
3. Transfer the inactivated sample out of the biosafety cabinet to the general laboratory area, according to established SOP.

3.5 Sample Dilution and Incubation

1. Serially dilute the inactivated sample in sample dilution buffer, add 100 μL of the mixture into the wells in duplicates.
2. Gently shake the plate for 2 min to mix well, and then incubate at 37 °C for 30 min while being covered with an adhesive plastic cover.
3. Discard the adhesive plastic cover and remove the solution.
4. Wash the plate with 300 μL of washing buffer per well for five times using an automatic microtiter plate washer.

3.6 Dilution of Secondary Antibody and Incubation

1. Dilute the secondary detection antibody (MAb 7C4 conjugated with HRP) in enzyme dilution buffer immediately before use.
2. Add 100 μL of the diluted detection antibody to each well using an 8-channel pipette.
3. Cover the plate with an adhesive plastic cover and incubate at 37 °C for 30 min.
4. Discard the adhesive plastic cover and remove the solution.
5. Wash the plate with 300 μL of washing buffer per well for five times using an automatic microtiter plate washer.

3.7 Detection and Readout

1. Add 100 μL of 3,3′,5,5′-Tetramethylbenzidine (TMB) substrate solution to each well (*see* **Note 5**).
2. Cover the plate with aluminum foil to protect from light, incubate for 10 min at room temperature.
3. Add 50 μL of stop solution to each well to stop the reaction.
4. Read the plate using an automatic plate reader at wavelength 450 nm.
5. Analyze the data by using the predetermined cutoff value.

4 Notes

1. An automated microtiter plate washer-dispenser would be a good addition to the workflow as the washing steps can be performed in a shorter amount of time and with greater consistency. But multichannel or even single-channel pipettes can be used instead.
2. The actual dilution of antibodies depends on the batch and quality of the MAbs used. Evaluations and characterization to determine the specificity and sensitivity have to be performed to establish the optimal dilution for the highest signal-to-noise ratio.
3. Prevent the microtiter plates to dry up by placing the plates in a box with some moist tissue paper laying under while storing in an incubator or on a rack in a warm water bath.

4. There are many viral lysis buffers available for purchase from bio-reagents vendors, e.g., buffer AL from Qiagen. Readers could request samples and perform their own testing on the conditions required to efficiently inactivate MERS-CoV.
5. TMB solutions are normally purchased from bio-reagents vendors at ready-to-use dilations, follow manufacturer's instructions.

Acknowledgments

This work is partly supported by the Theme-based Research Scheme (Project No. T11/707/15), University Grant Committee; Consultancy Service for Enhancing Laboratory Surveillance of Emerging Infectious Diseases and Research Capability on Antimicrobial Resistance for Department of Health, HKSAR Government; and University Development Fund, The University of Hong Kong.

References

1. (WHO), W.H.O., MERS situation update, July 2018. 2018, World Health Organization: World Health Organization, 20 Avenue Appia, 1211 Geneva 27, Switzerland. p. 1
2. (WHO), W.H.O., Laboratory Testing for Middle East Respiratory Syndrome Coronavirus. 2018 (January 2018): p. 8
3. Chen Y et al (2015) A sensitive and specific antigen detection assay for Middle East respiratory syndrome coronavirus. Emerg Microbes Infect 4(4):e26
4. Corman VM et al (2012) Detection of a novel human coronavirus by real-time reverse-transcription polymerase chain reaction. Euro Surveill 17(39):20285
5. Corman VM et al (2012) Assays for laboratory confirmation of novel human coronavirus (hCoV-EMC) infections. Euro Surveill 17 (49):20334
6. Muller MA et al (2015) Presence of Middle East respiratory syndrome coronavirus antibodies in Saudi Arabia: a nationwide, cross-sectional, serological study. Lancet Infect Dis 15(6):629
7. Perera RA et al (2013) Seroepidemiology for MERS coronavirus using microneutralisation and pseudoparticle virus neutralisation assays reveal a high prevalence of antibody in dromedary camels in Egypt, June 2013. Euro Surveill 18(36):pii=20574
8. Saeidnia S, Abdollahi M (2013) Are other fluorescent tags used instead of ethidium bromide safer? Daru 21(1):71
9. Drosten C et al (2013) Clinical features and virological analysis of a case of Middle East respiratory syndrome coronavirus infection. Lancet Infect Dis 13(9):745–751
10. Kapoor M et al (2014) Clinical and laboratory findings of the first imported case of Middle East respiratory syndrome coronavirus to the United States. Clin Infect Dis 59 (11):1511–1518
11. Alagaili AN et al (2014) Middle East respiratory syndrome coronavirus infection in dromedary camels in Saudi Arabia. MBio 5(2): e00884–e00814
12. Chen Y et al (2016) A highly specific rapid antigen detection assay for on-site diagnosis of MERS. J Infect 73(1):82–84
13. Woo PCY et al (2018) Rapid detection of MERS coronavirus-like viruses in bats: potelntial for tracking MERS coronavirus transmission and animal origin. Emerg Microbes Infect 7(1):18
14. Lau SK et al (2013) Genetic characterization of Betacoronavirus lineage C viruses in bats reveals marked sequence divergence in the spike protein of pipistrellus bat coronavirus HKU5 in Japanese pipistrelle: implications for the origin of the novel Middle East respiratory

syndrome coronavirus. J Virol 87(15): 8638–8650

15. Yuan Q et al (2011) Differential diagnosis of pandemic (H1N1) 2009 infection by detection of haemagglutinin with an enzyme-linked immunoassay. Clin Microbiol Infect 17 (10):1574–1580
16. Lau SK et al (2005) SARS coronavirus detection methods. Emerg Infect Dis 11 (7):1108–1111
17. Lau SK et al (2004) Detection of severe acute respiratory syndrome (SARS) coronavirus nucleocapsid protein in sars patients by enzyme-linked immunosorbent assay. J Clin Microbiol 42(7):2884–2889
18. Chan JF et al (2013) Differential cell line susceptibility to the emerging novel human betacoronavirus 2c EMC/2012: implications for disease pathogenesis and clinical manifestation. J Infect Dis 207(11):1743–1752
19. Chan KH et al (2013) Cross-reactive antibodies in convalescent SARS patients' sera against the emerging novel human coronavirus EMC (2012) by both immunofluorescent and neutralizing antibody tests. J Infect 67(2): 130–140

Chapter 8

Quantification of the Middle East Respiratory Syndrome-Coronavirus RNA in Tissues by Quantitative Real-Time RT-PCR

Abdullah Algaissi, Anurodh S. Agrawal, Anwar M. Hashem, and Chien-Te K. Tseng

Abstract

Since the emergence of the Middle East respiratory syndrome-coronavirus (MERS-CoV) in 2012, more than 2280 confirmed human infections and 800 associated deaths had been reported to the World Health Organization. MERS-CoV is a single-stranded RNA virus that belongs to the *Coronaviridae* family. MERS-CoV infection leads to a variety of clinical outcomes in humans ranging from asymptomatic and mild infection to severe acute lung injury and multi-organ failure and death. To study the pathogenesis of MERS-CoV infection and development of medical countermeasures (MCMs) for MERS, a number of genetically modified mouse models have been developed, including various versions of transgenic mice expressing the human DPP4 viral receptor. Tracking and quantifying viral infection, among others, in permissive hosts is a key endpoint for studying MERS pathogenesis and evaluating the efficacy of selected MCMs developed for MERS. In addition to quantifying infectious progeny virus which requires high-containment biosafety level (BSL)-3 laboratory, here we outlined an established real-time quantitative RT-PCR (RT-qPCR)-based procedure to unequivocally quantify MERS-CoV-specific RNAs within the lungs of infected human DPP4 (hDPP4, transgenic (hDPP4 Tg) mice under a standard BSL-2 laboratory.

Key words MERS-CoV, RT-qPCR, Animal models

1 Introduction

Middle East respiratory syndrome-coronavirus (MERS-CoV) is an emerging coronavirus that was first identified in Saudi Arabia in 2012 [1]. Since its emergence, MERS-CoV has infected more than 2280 individuals with over 800 deaths in 27 countries around the world, with the majority of the infections occurring in Saudi Arabia. MERS-CoV is classified as a lineage C betacoronavirus (Beta-CoV). Beta-CoV is one of four genera of the coronaviruses of the subfamily *Coronavirinae* in the family *Coronaviridae* [2]. In addition to MERS-CoV, the Beta-CoV genus also contains the SARS-CoV (lineage B), and other human coronaviruses such as OC43 and

Rahul Vijay (ed.), *MERS Coronavirus: Methods and Protocols*, Methods in Molecular Biology, vol. 2099, https://doi.org/10.1007/978-1-0716-0211-9_8,

HKU1 (lineage A) [2]. Coronaviruses are enveloped RNA viruses with a large positive-sense, single-stranded genome that ranges in size from 28 to 32 kbp and characterized by having crown-like projections on the virus particles [3]. The genome of the MERS-CoV is about 30.1 kbp long and contains 11 open reading frames (ORFs). Like other coronaviruses, the first two-thirds of the MERS-CoV genome contains two overlapping ORFs, ORF1a and ORF1b, that encode the viral replicase-transcriptase complex, the nonstructural proteins 1–16 (nsp1–16). The remaining one-third of the genome comprises ORFs that encodes the structural proteins, which include the spike (S), envelope (E), membrane (M), and nucleocapsid (N) proteins. In addition to these structural and nonstructural proteins shared by other coronaviruses, the MERS-CoV genome also contains several ORFs coding for accessory proteins, namely ORF3, ORF4a, and ORF4b [4].

Because of its high mortality rate and its potential to spread worldwide, it is important to study MERS-CoV pathogenesis in animal models. Mice are the most common and accessible laboratory animal species used for biomedical research in general. However, due to the disparity of two amino acids in the viral receptor—dipeptidyl peptidase 4 (DPP4)—from the human sequence that are critical for efficient binding to the receptor binding domain (RBD) of MERS-CoV, wild-type mice are not naturally permissive to MERS-CoV [5, 6]. We have recently developed a transgenic (Tg) mouse model, expressing the human DPP4, that is highly permissive to MERS-CoV infection and disease [7, 8]. The quantification of MERS-CoV RNA in specimens is important for studying MERS pathogenesis, evaluating the efficacy of selected MCMs and diagnostics. In this protocol, we outlined a stepwise RT-qPCR-based protocol to unequivocal quantify MERS-CoV-specific RNAs within the lungs of infected hDPP4 Tg mice.

2 Materials

2.1 General Materials

1. Titrated MERS-CoV (EMC-2012 strain).
2. Vero E6 cells (ATCC® CRL-1586).
3. Minimal Essential Media (MEM): Minimal Essential Media (MEM) supplemented with 2% heat inactivated FCS, 1% L-glutamine, and 1% penicillin/streptomycin, referred to as M-2.
4. 2.0 mL screw cap tubes.
5. Polypropylene microcentrifuge tubes.
6. RNAlater RNA stabilization solution.
7. Tissue Homogenizer.

Table 1
Primers and probes for RT-qPCR

Primers and probe for MERS-CoV upE (see **Note 1**)	
Primer forward (5′-3′)	GCAACGCGC GAT TCAGTT
Primer reverse (5′-3′)	GCCTCTACACGGGACCCATA
Probe (5′-3′)	FAM/CTCTTCACATAATCGCCCCGAGCTCG/TAMRA
Primers and probe for endogenous control (mouse β-actin)	
Primer forward (5′-3′)	CTGGATGGCTACGTACATGG
Primer reverse (5′-3′)	ACCTTCACAATGAGCTGCG
Probe (5′-3′)	FAM/TCTGGGTCATCTTTTCACGGTTGGC/TAMRA

2.2 RNA Extraction

1. TRIzol® reagent.
2. Stainless steel beads, 5 mm.
3. Chloroform.
4. Isopropanol.
5. Ethanol, 75%.
6. Refrigerated centrifuge and rotor capable of atleast12,000 × g speed.
7. RNA storage solution.
8. Method to quantify RNA concentration in specimens (SpectraMax i3 Multi-mode Microplate reader from Molecular Device or other machines).

2.3 Viral RNA Quantification

1. TaqMan® primers/probes set for upE gene of MERS-CoV and endogenous host control (Table 1).
2. Superscript III One-Step RT-PCR kit.
3. Optical 96-well PCR plate or PCR tubes.
4. RT-qPCR machine.

3 Method

Biosafety: MERS-CoV is a biosafety level 3 (BSL3) pathogen. Thus, all work involves infectious MERS-CoV should be handled in a BSL3 facility following the institutional guidelines and regulations such as wearing the proper PPE and having the proper training to work at BSL3 and animal BSL3 laboratories. Since RNA is very unstable, make sure to use RNAse-free tubes and reagents. Always make sure to clean all the pipettes and bench to using any preferred reagent to remove RNAse.

3.1 Infection (Intranasal Route)

1. Prepare a virus inoculum in M-2 media as a concentration of 1×10^3 $TCID_{50}$ in 50 μl (equals to 100 LD_{50} in our Tg mouse model) [8]. Keep the virus on ice. Mock-infected mice should be included as a control to calculate the relative expression of MERS-CoV RNA compared to control.
2. Anesthetize mice using isoflurane vaporizer in an induction chamber (other approved methods like injectable anesthesia can also be used).
3. Once the mice are completely asleep, hold vertically and slowly deliver the virus inoculum (50 μl) into their nostrils. Make sure the solution gets entirely into the nose and not swallowed through the mouth.
4. Place the mice back in the cage until the desired time point to measure virus titer in the lung.

3.2 Tissue Collection and Homogenization

All procedures of RNA tissue homogenization and extraction should be done in BSL3 laboratory.

1. At the desired time point, euthanize the mice using CO_2 or other approved methods. In our Tg mice, we detect the highest viral titer in the lung at day 2 and 3 post infection (*see* **Note 2**).
2. Using standard necropsy technique, collect a piece of lung tissues (one-quarter of a lung) immediately into a tube containing RNAlater solution. The volume of the RNAlater solution should be at least ten times the size of the tissue.
3. Store the tissue at 4 °C until homogenization.
4. After at least 24 h, weight tissues and transfer into a 2.0 mL screw cap tube containing 1 mL TRIzol reagent and two stainless steel beads. The sample size should not exceed 10% of the volume of TRIzol reagent used for homogenization.
5. Under the biosafety cabinet, homogenize tissue using automated tissue homogenizer (2× for 60 s at 25 strokes/s).
6. Spin down for 1 min at 5000 × *g* to pellet tissue debris and collect supernatant that contains total RNA into fresh screw cap tubes.
7. Incubate at room temperature for 2–5 min after homogenization. You can proceed with total RNA isolation immediately or store the homogenized sample at −80 °C.

3.3 Total RNA Extraction

1. Add 0.2 mL chloroform to the tube for each 1 mL TRIzol reagent used for homogenization.
2. Shake vigorously by hand for 15–30 s and incubate at room temperature for 5 min.
3. Centrifuge at 12,000 × *g* for 20 min at 4 °C.
4. Carefully remove the upper aqueous phase, which contains the total RNA, and place into a fresh microcentrifuge tube.

5. Add 0.5 mL of isopropanol to the tube containing total RNA and incubate at room temperature for 10 min (or at −20 °C for 1 h).
6. Centrifuge at 12,000 × *g* for 10 min at 4 °C.
7. Remove the supernatant leaving only the RNA that appears as a gel-like pellet on the side of the tube.
8. Wash the pellet with at least 1 mL of 75% ethanol per 1 mL TRIzol used.
9. Centrifuge at 12,000 × *g* for 5 min at 4 °C and discard the supernatant.
10. Repeat the wash one more time (**steps 8** and **9**) to ensure complete removal of guanidine salt present in sample.
11. Remove supernatant and dry the pellet at room temperature for no more than 10 min (do not let the pellet dry completely).
12. Suspend the pellet in 150–200 μL of RNA storage solution.
13. Measure the concentration of RNA in a SpectraMax i3 Multi-mode Microplate reader or equivalent.
14. The expected total RNA yield from 1 mg lung tissue is 5–10 μg.
15. Keep RNA at −80 °C until RT-qPCR analysis or on ice if proceeding immediately.

3.4 Quantitative Real-Time RT-PCR (RT-qPCR) and Calculation of Relative Copy Number of MERS-CoV RNA

RT-qPCR to quantify MERS-CoV RNA is performed in triplicate using the Superscript III One-Step RT-PCR kit (Invitrogen) with MERS-CoV-specific primers and probes (Table 1):

1. Set up a 25 μL one-step RT-qPCR reaction in an optical 96-well plate or on PCR tubes on ice: SuperScript III RT/Platinum Taq mix (1 μL), 2× reaction (12 μL), forward primer 10 μM (1 μL), reverse primer 10 μM (1 μL), fluorogenic probe 10 μM (1 μL), 1 μg of the total RNA (≤5 μL), and up to 25 μL of RNAse/DNase-free water (*see* **Notes 3** and **4**).
2. Seal or cap the PCR reaction tube/plate and gently mix.
3. Centrifuge the plate or the tubes for 1 min at 500 × *g* to settle down any droplets on the inner sides of the wells.
4. Place the PCR tubes/plate in a preheated real-time PCR machine and run RT-qPCR reaction using the conditions recommended by the manufacturer as shown in Table 2.
5. Calculate the relative copy number of MERS-CoV RNA normalized to the endogenous control (mouse β-actin) using the standard threshold cycle (ΔΔCt) as follows:

   ΔCt 1 (MERS-CoV infected) = Ct MERS-COV upE − Ct β-actin.

Table 2
RT-qPCR program conditions

Step	Temperature (°C)	Time	Cycles
Reverse transcription	50	30 min	1
Initial denaturing	95	2 min	1
Denaturation Annealing/extension	95 60	15 s 30 s	40

ΔCt 2 (Control (Mock-infected)) = Ct MERS-COV upE − Ct β-actin.

ΔΔCt = ΔCt 1 − ΔCt 2.

The relative expression of MERS-CoV upE RNA in the sample compared to control = 2- ΔΔCt (*see* **Note 5**).

3.5 Standard Curve to Quantify MERS-CoV RNA as $TCID_{50}$Eq/Gram of Tissue

To determine the amount of the viral RNA load as $TCID_{50}$ eq/gram of tissue, a standard curve need to be generated.

1. Collect lung tissues from uninfected hDPP4 Tg mice and place them in a vial containing 1 mL RNAlater solution before extracting total RNAs as described in Subheading 3.2. This sample will be spiked with viral inoculum to serve as a positive control for standard curve preparation.
2. Weigh and homogenize the collected lung tissue as described.
3. To make the positive control standards, divide the homogenized sample into five equal aliquots, each containing 1 mL of Trizol reagent (*see* **Note 6**).
4. Spike the aliquots with different dilutions of MERS-CoV starting from 10^5 $TCID_{50}$/mL until 10^1 $TCID_{50}$/mL (*see* **Note 6**).
5. Isolate spiked RNA from these samples as described in Subheading 3.3.
6. Use 1 μg of spiked RNA from each standard to perform one-step real-time RT-qPCR in triplicate using Superscript III One-Step RT-PCR kit as described in Subheading 3.4.
7. Determine the mean Ct value of each standard dilution, which corresponds to the viral titers.
8. Construct a 5-point standard curve by plotting the obtained mean Ct values against the titers of MERS-CoV used for spiking as $TCID_{50}$/mL (Fig. 1).
9. Calculate the $TCID_{50}$ eq/gram of tissue of your sample using the established standard curve.

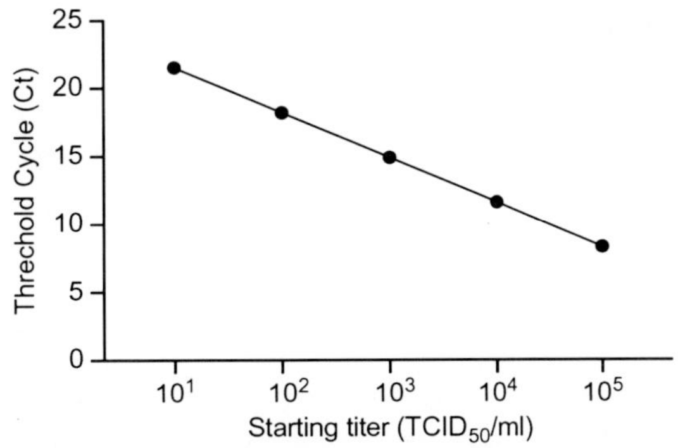

Fig. 1 Standard curve of the tenfold dilution (10^5 to 10^1) of MERS-CoV by real-time RT-qPCR: Lung tissue were isolated from uninfected hDPP4 Tg mouse, and homogenized. For RNA isolation, homogenized lung tissues sample were divided into five equal parts and spiked with different dilutions of $TCID_{50}$/mL of stock MERS-CoV in the range from 10^5 $TCID_{50}$/mL to 10^1 $TCID_{50}$/mL. Standard curve was generated by RT-qPCR technique using real-time PCR machine by loading uniform amount of RNA in triplicate. Total of 5 standards were constructed by plotting mean Ct values over titers of infectious MERS-CoV in the samples

4 Notes

1. Probes with different fluorescent dyes can be used based on the used instrument and the manufacturer's instructions.
2. We focus here on lung tissues because the lung is the target organ for MERS-CoV infection. However, viral RNA can be detected in other organs such as heart, brain, and spleen in low titers [8]. This protocol can be followed to detect viral RNA in these tissues as well.
3. In the protocol, we use one-step RT-qPCR without the need for separate reaction for cDNA synthesis, which is quick and simple. However, creating a cDNA first can be done and used for qPCR, especially if there is a need to stock cDNA to quantify other targets.
4. For multiple samples, prepare a master mix and add an appropriate volume to the plate wells, followed by adding the RNA template. Master mix preparation is crucial to reduce pipetting errors.
5. Alternatively, the viral RNA titer can be represented as TCID50 eq/gram of tissue as detailed in Subheading 3.5.
6. More aliquots could be prepared to make higher range standard curve by spiking with higher concentrations of MERS-CoV.

Acknowledgment

This work was supported by King Abdulaziz City for Science and Technology (KACST) through the MERS-CoV research grant program (number 09-1 to AMH and CKT), which is a part of the Targeted Research Program.

References

1. Zaki AM, van Boheemen S, Bestebroer TM et al (2012) Isolation of a novel coronavirus from a man with pneumonia in Saudi Arabia. N Engl J Med 367:1814–1820. https://doi.org/10.1056/NEJMoa1211721
2. Chan JFW, Lau SKP, To KKW et al (2015) Middle East respiratory syndrome coronavirus: another zoonotic betacoronavirus causing SARS-like disease. Clin Microbiol Rev 28:465–522. https://doi.org/10.1128/CMR.00102-14
3. Fehr AR, Perlman S (2015) Coronaviruses: an overview of their replication and pathogenesis. Methods Mol Biol 1282:1–23. https://doi.org/10.1007/978-1-4939-2438-7_1
4. de Wit E, van Doremalen N, Falzarano D et al (2016) SARS and MERS: recent insights into emerging coronaviruses. Nat Rev Microbiol 14:523–534. https://doi.org/10.1038/nrmicro.2016.81
5. Coleman CM, Matthews KL, Goicochea L et al (2014) Wild-type and innate immune-deficient mice are not susceptible to the Middle East respiratory syndrome coronavirus. J Gen Virol 95:408–412. https://doi.org/10.1099/vir.0.060640-0
6. Cockrell AS, Peck KM, Yount BL et al (2014) Mouse Dipeptidyl peptidase 4 is not a functional receptor for Middle East respiratory syndrome coronavirus infection. J Virol 88:5195–5199. https://doi.org/10.1128/JVI.03764-13
7. Agrawal AS, Garron T, Tao X et al (2015) Generation of a transgenic mouse model of Middle East respiratory syndrome coronavirus infection and disease. J Virol 89:3659–3670. https://doi.org/10.1128/JVI.03427-14
8. Tao X, Garron T, Agrawal AS et al (2015) Characterization and demonstration of value of a lethal mouse model of Middle East respiratory syndrome coronavirus infection and disease. J Virol 90:57–67. https://doi.org/10.1128/JVI.02009-15

Chapter 9

Evaluation of MERS-CoV Neutralizing Antibodies in Sera Using Live Virus Microneutralization Assay

Abdullah Algaissi and Anwar M. Hashem

Abstract

The microneutralization (MN) assay is a standard and important technique in virology, immunology, and epidemiology. It is a highly specific and sensitive assay for evaluating virus-specific neutralizing antibodies (nAbs) in human and animal sera. It provides the most precise answer to whether or not an individual or animal has antibodies that can neutralize or inhibit the infectivity of a specific virus strain. However, using live virus-based MN assay might require working under high containment facilities especially when dealing with high-risk pathogens such as the Middle East respiratory syndrome-coronavirus (MERS-CoV). In this chapter, we describe the isolation, amplification, and titration of MERS-CoV, as well as detailed MN assay to measure nAb levels in sera from different mammalian species.

Key words MERS-CoV, Neutralizing antibodies, Microneutralization

1 Introduction

The Middle East respiratory syndrome-coronavirus (MERS-CoV) is a novel zoonotic β-coronavirus that was first identified in Saudi Arabia in 2012 [1]. Epidemiological evidence suggests that dromedary camels are the main zoonotic source of MERS-CoV [2, 3]. MERS-CoV causes a wide range of manifestations ranging from asymptomatic infections to mild or severe respiratory disease. Detection of anti-MERS-CoV antibodies (Abs) in humans and/or animals represents a valuable tool in diagnostics as well as epidemiological, virological, and immunological studies including evaluation of vaccine immunogenicity [4–8]. Several serological assays have been developed and used for MERS-CoV, including ELISA-based assays, immunofluorescence assays, protein microarrays, and pseudovirus-based neutralization assays [9–17]. However, most of these assays pose several drawbacks and limitations such as low specificity and sensitivity, need for expensive and special equipment and reagents, and/or highly trained technical staff, which could limit their use.

Rahul Vijay (ed.), *MERS Coronavirus: Methods and Protocols*, Methods in Molecular Biology, vol. 2099,
https://doi.org/10.1007/978-1-0716-0211-9_9,

On the other hand, live virus-based microneutralization (MN) assay is a highly sensitive and specific technique used for the quantitation of virus-specific neutralizing antibodies (nAbs) to a given virus in mammalian sera as well as the evaluation of antiviral activities of small molecules and biologics. The assay has several advantages in detecting nAbs against MERS-CoV. It can precisely detect virus-specific nAbs in human and animal sera without the need for specific reagents or equipment, it can be carried out readily once the virus is isolated, and it can overcome strain-specific antigenic changes. The protocol presented in this chapter consists of four major steps, including MERS-CoV isolation, amplification, titration, and neutralization. The MN assay described here is suitable to quantitatively measure the titer of nAbs in sera from different mammalian species.

2 Materials

2.1 General Materials

1. Vero E6 cells.
2. 37 °C water bath.
3. Biosafety cabinet.
4. 37 °C incubator with 5% CO_2.
5. Low-speed centrifuge.
6. Inverted microscope.
7. Cell counter or hemocytometer.
8. M-10 media: Dulbecco's modification of Eagle medium (DMEM), 10% heat-inactivated fetal bovine serum (FBS), 1% L-glutamine, and 1% penicillin/streptomycin.
9. M-2: Dulbecco's modification of Eagle medium (DMEM), 2% heat-inactivated fetal bovine serum (FBS), 1% L-glutamine, and 1% penicillin/streptomycin.
10. 70% Ethanol for decontamination of laminar flow biosafety cabinet and objects brought into the hood.
11. Sterile serological pipettes.
12. Sterile DPBS without calcium or magnesium.
13. Sterile 1× trypsin-EDTA in DPBS without calcium or magnesium.
14. Sterile T25, T75, and T175 tissue culture flasks with vented caps.
15. Sterile 15 mL falcon tubes.
16. Sterile 1.5 mL tubes.
17. 1 mL pipette.
18. Multichannel pipette.

19. Sterile disposable aerosol-resistant filtered tips.
20. Sterile 0.22 μm, γ-irradiated syringe filters.

2.2 MERS-CoV Isolation, Amplification, Titration, and Microneutralization

MERS-CoV is a biosafety level 3 (BSL3) pathogen. Thus, all work involving infectious MERS-CoV should be handled in a BSL3 facility following the institutional guidelines and regulations such as wearing the proper PPE and having the proper training to work at a BSL3 facility.

1. Confluent Vero E6 cells cultured in M-10 in a T75 tissue culture flask.
2. Sterile 96-well tissue culture plates.
3. Sterile U-shaped 96-well plates.
4. MERS-CoV-positive sample: filtered using Sterile 0.22 μm, γ-irradiated syringe filters (to be used for virus isolation).
5. Isolated MERS-CoV (to be titrated or amplified).
6. Titrated MERS-CoV (for microneutralization assay).
7. Test serum samples.

3 Methods

3.1 Continuous Culture of Vero E6 Cells

Proper aseptic techniques should be used, and all equipment and solutions to be used with cells must be sterile. All cell culture incubations should be performed in a humidified 37 °C incubator with 5% CO_2, and all solutions should be pre-warmed to room temperature or 37 °C before use with cells.

1. Remove a cryovial of frozen Vero E6 cells from liquid nitrogen storage and quickly thaw the cells for <1 min by gently swirling in a 37 °C water bath until there is just a small bit of ice left in the vial (*see* **Note 1**).
2. Transfer the vial into a laminar flow biosafety cabinet and quickly disinfect the outside of the vial with 70% ethanol.
3. Open the vial and transfer all the volume to a sterile 15 mL falcon tube containing 10 mL of pre-warmed M-10 (*see* **Note 2**).
4. Centrifuge the cell suspension at approximately 200–500 × *g* for 5 min at room temperature.
5. Aseptically decant the supernatant without disturbing the cell pellet (*see* **Note 3**).
6. Gently re-suspend the cells in 10 mL of pre-warmed M-10 (*see* **Note 4**).
7. Transfer the cell suspension to T75 tissue culture flask using sterile 10 mL serological pipette.
8. Incubate the flask in 37 °C incubator with 5% CO_2.

9. Monitor cells under an inverted microscope daily or every other day and change media every 3–4 days if necessary (*see* **Note 5**).
10. When cells reach >90–95% confluency, passage cells into new tissue culture flasks at 1:5 to 1:10 split ratio (*see* **Notes 6** and 7).
11. Maintain cells in continuous culture and passage them as needed for at least 2–3 passages after removal from long-term storage and before use.

3.2 Isolation and Amplification of MERS-CoV

1. Harvest confluent Vero E6 cells from a T75 tissue culture flask using standard trypsinization procedure (*see* **Note** 7).
2. Count cells using cell counter or hemocytometer and prepare a cell suspension of 5×10^5 cells/mL in pre-warmed M-10.
3. Seed 5 mL (~2.5×10^6 cells) or 20 mL (~1×10^7 cells) of the cell suspension into a T25 or T175 tissue culture flask, respectively, so that they are 90–95% confluent the next day (*see* **Note 8**).
4. Incubate the flasks in 37 °C incubator with 5% CO_2 for overnight.
5. Next day, change the media by removing old media and add 2 mL or 5 mL of fresh pre-warmed M-2 into a T25 or T175 tissue culture flask, respectively.
6. If using positive MERS-CoV sample, filter sterilize samples using Sterile 0.22 μm γ-irradiated syringe filters before inoculation onto Vero E6 cells (*see* **Note 9**).
7. Add 0.5–1 mL of isolated MERS-CoV or filtered positive sample to the cells (*see* **Note 10**).
8. Distribute the virus evenly over the cells and incubate for 1 h in 37 °C incubator with 5% CO_2.
9. Make up the final volume of media to 5 mL or 20 mL in T25 or T175 tissue culture flasks, respectively.
10. Incubate the flask in 37 °C incubator with 5% CO_2 for 2–3 days or until significant cytopathic effect (CPE) is observed (Fig. 1).
11. Check the flask daily post-infection (*see* **Note 11**).
12. When CPE is >50%, collect supernatant from the flask and centrifuge at 500 × *g* for 5 min to remove cellular debris (Fig. 1).
13. Aliquot collected clarified supernatant in 100 μL or 1 mL aliquots in sterile 1.5 mL tubes and store at −80 °C (*see* **Note 12**).

3.3 Titration of MERS-CoV by Tissue Culture Infective Dose 50 ($TCID_{50}$)

1. Harvest confluent Vero E6 cells from the T75 tissue culture flask using standard trypsinization procedure (*see* **Note** 7).
2. Count the cells using cell counter or hemocytometer and prepare a cell suspension of 1×10^5 cells/mL in pre-warmed M-10. Re-suspend 1×10^6 cells in 10 mL per 96-well plate.

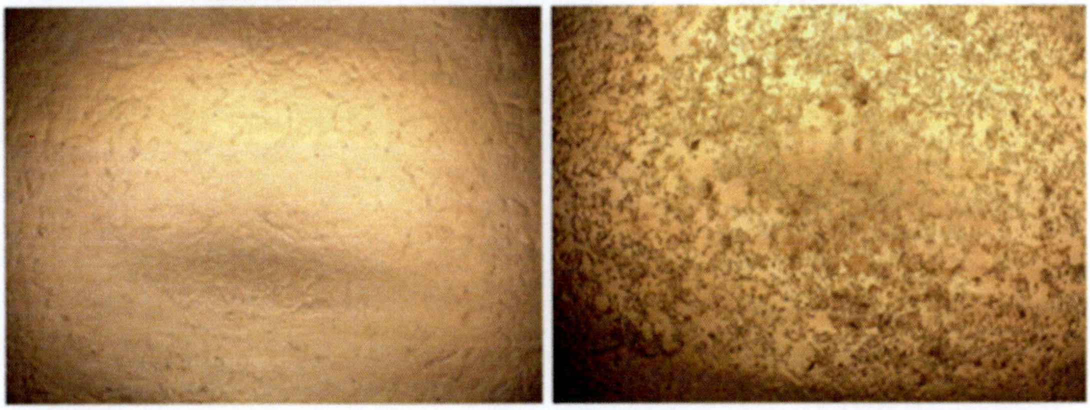

Fig. 1 Example of cytopathic effect (CPE) observed in Vero E6 cells infected with MERS-CoV 3 days after inoculation

3. Seed 1 × 10^4 Vero E6 cells (100 μL) per well into sterile 96-well tissue culture plate so that they are 90–95% confluent the next day (*see* **Note 8**).
4. Incubate the plate in 37 °C incubator with 5% CO_2 for overnight.
5. Next day, in a new sterile U-shaped 96-well plate, add 135 μL pre-warmed M-2 to all wells (Fig. 2).
6. Add 15 μL of MERS-CoV per well in all wells of column 1 to have 1:10 dilution (Fig. 2).
7. Perform tenfold serial (log10) dilution by transferring 15 μL progressively from column to column (Fig. 2) using a multi-channel pipette (*see* **Note 13**).
8. During each dilution step, mix well by pipetting eight times up and down (*see* **Note 14**).
9. Discard the final 15 μL after column 11 (i.e., wells in column 12 should not contain virus, cell control (CC)).
10. Remove the 96-well tissue culture plate containing confluent Vero E6 cells and aspirate the media (*see* **Note 15**).
11. Transfer 100 μL from the U-shaped 96-well plate containing diluted MERS-CoV sample to the cells in each corresponding well in the 96-well tissue culture plate using a multichannel pipette (*see* **Note 14**).
12. Incubate the 96-well tissue culture plate in 37 °C incubator with 5% CO_2 for 3 days (*see* **Note 11**).
13. After incubation, observe plate under an inverted microscope and score wells as positive for MERS-CoV (i.e., CPE) or negative for MERS-CoV (i.e., cells are intact and no CPE) (*see* **Note 16**).
14. Calculate $TCID_{50}$ using Reed–Muench formula [18].

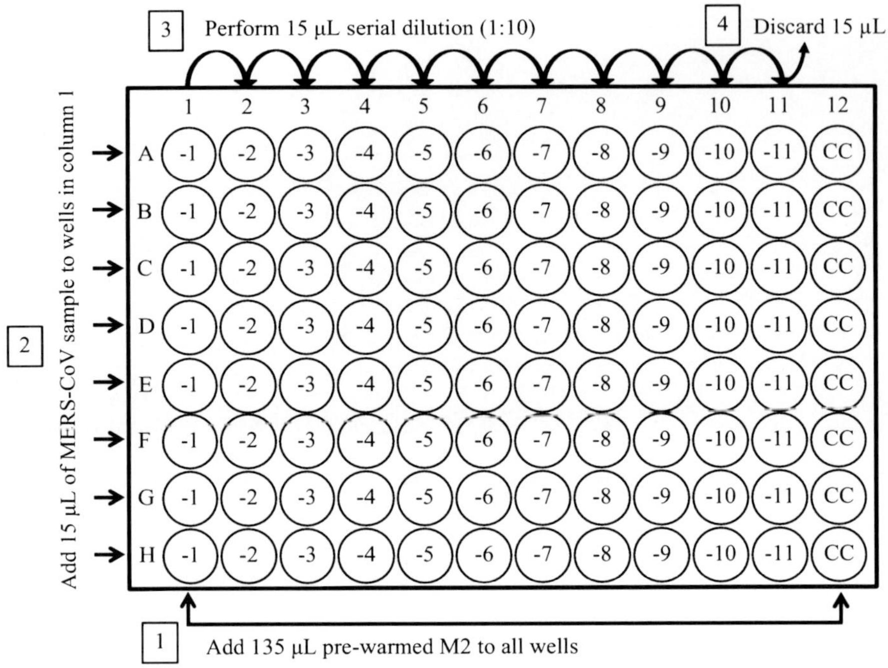

Fig. 2 Schematic representation of preparation of virus dilution for MERS-CoV titration by $TCID_{50}$. The sequential steps of plate preparation are indicated by numbered boxes. After completion, transfer 100 μL from the 96-well U-bottom plate containing diluted MERS-CoV sample to the cells in each corresponding well in the 96-well tissue culture plate using a multichannel pipette. CC is cell control wells

3.4 MERS-CoV Microneutralization Assay

1. Harvest confluent Vero E6 cells from the T75 tissue culture flask using standard trypsinization procedure (*see* **Note 7**).
2. Count the cells using cell counter or hemocytometer and prepare a cell suspension of 1×10^5 cells/mL in pre-warmed M-10. Re-suspend 1×10^6 cells in 10 mL per 96-well plate.
3. Seed 1×10^4 Vero E6 cells (100 μL) per well into sterile 96-well tissue culture plate so that they are 90–95% confluent the next day (*see* **Note 8**).
4. Incubate the plate in 37 °C incubator with 5% CO_2 overnight.
5. Next day, heat-inactivate test sera to be used for virus microneutralization by incubation for 30 min at 56 °C.
6. In a new sterile U-shaped 96-well plate, add 60 μL pre-warmed M-2 to all wells (Fig. 3).
7. Add an additional 48 μL pre-warmed M-2 to wells A1–A10 in row A (Fig. 3).
8. Add 12 μL heat-inactivated serum per well in wells A1–A10 in row A to have 1:10 dilution (Fig. 3). Do not add serum to A11 and A12 (*see* **Note 17**).

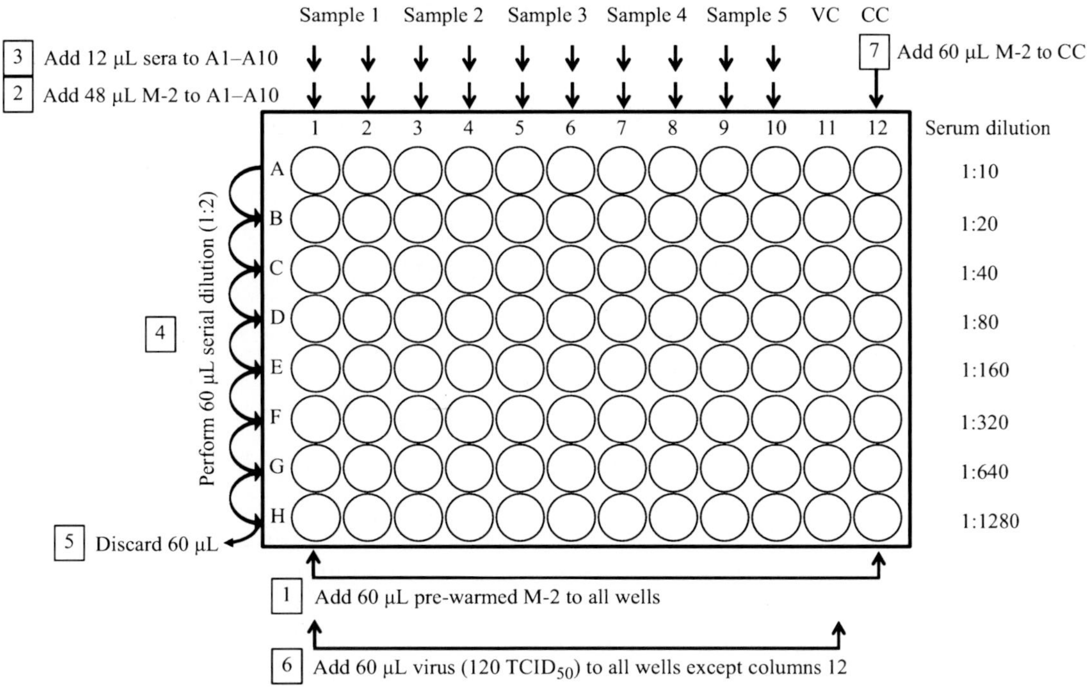

Fig. 3 Plate preparation for MERS-CoV microneutralization assay. The preparation steps for MERS-CoV microneutralization assay plate are indicated by numbered boxes. After completion, incubate the serum-virus mixtures for 1 h in 37 °C incubator with 5% CO_2. Then, transfer 100 μL from the U-shaped 96-well plate containing diluted MERS-CoV sample to the cells in each corresponding well in the 96-well tissue culture plate using a multichannel pipette. VC is virus control and CC is cell control wells

9. Perform twofold serial dilutions on added serum samples by transferring 60 μL progressively from row to row (i.e., A1 to B1; B1 to C1; etc. up to G1 to H1) using a multichannel pipette (Fig. 3).
10. During each dilution step mix well by pipetting eight times up and down (*see* **Note 14**).
11. Discard the final 60 μL after row H.
12. Prepare virus suspension in pre-warmed M-2 so that 60 μL contains 120 $TCID_{50}$ (i.e., 2×10^3 $TCID_{50}$/mL). Approximately 6 mL/plate is needed (*see* **Note 18**).
13. Add 60 μL diluted virus to all wells except wells in columns 12 (CC wells).
14. Add 60 μL pre-warmed M-2 to all CC wells (i.e., columns 12).
15. Incubate the serum-virus mixtures for 1 h in 37 °C incubator with 5% CO_2.
16. Remove the 96-well tissue culture plate containing confluent Vero E6 cells and aspirate the media (*see* **Note 15**).

17. Transfer 100 μL from the U-shaped 96-well plate to the cells in each corresponding well in the 96-well tissue culture plate using a multichannel pipette (*see* **Note 14**).
18. Incubate the 96-well tissue culture plate in 37 °C incubator with 5% CO_2 for 3 days (*see* **Note 11**).
19. Calculate MN_{50} or MN_{100} titers of each serum sample as the highest serum dilution that completely protect the cells from CPE in half or all wells, respectively.

4 Notes

1. The water bath is a potential source of contamination. To reduce the risk of contamination, keep the O-ring and cap of the cryovial out of the water.
2. Frozen cell stocks contain dimethyl sulfoxide (DMSO), which is harmful to the cells, and it should be diluted and removed after thawing the cells and before transferring the cells to tissue culture flasks.
3. After centrifugation, check the clarity of the supernatant and visibility of a complete pellet.
4. Different volumes and culture vessels could be used. It is better to initiate Vero E6 cells culture in a T25 tissue culture flask. If using a T25 tissue culture flask, re-suspend the cells in 5 mL media, and if using T75 tissue culture flask, re-suspend the cells in 10 mL media.
5. Vero E6 cells recover slowly after freezing and may take more than a week before they are ready to be passaged. It may take 2–3 passages before the Vero E6 cells reach their normal growth rate.
6. It is important to monitor Vero E6 cells and to subculture them once confluent. Depending on the number of seeded cells and the size of the used flask, Vero E6 cells usually need to be passaged 2–3 times per week.
7. To harvest or maintain Vero E6 cells, remove media from the flask, wash the cell monolayer gently with 3–5 mL of sterile pre-warmed DPBS without calcium or magnesium, and discard the used washing solution. Add 2–5 mL pre-warmed 1× trypsin-EDTA in DPBS without calcium or magnesium to the cell monolayer and incubate for 5–10 min at 37 °C, 5% CO_2 to detach cells (incubation may vary, so check the cells every 2–3 min). After cells are detached, add 5 mL pre-warmed M-10 to the flask to inactivate trypsin activity, and collect detached cells in 15 mL sterile falcon tube. Make sure to centrifuge the collected cells and discard the supernatant.

Then, add new 1–2 mL pre-warmed M-10 and re-suspend the cells by pipetting up and down using 1 mL pipette to make a homogenous cell suspension.

8. Change the seeding density of cells when cells are under or over confluent.
9. Use, Sterile 0.22 μm, γ-irradiated syringe filters with small membrane diameter if sample volume is small for minimal sample loss.
10. Multiplicity of infection (MOI) of 0.001–0.1 could be used if titer is known. It is suggested to use small culture vessels (T25 tissue culture flask) for virus isolation from positive samples and larger culture vessels (T175 tissue culture flask) for virus amplification.
11. CPE could be strain specific, and it depends on the strain and starting titer of the seed virus.
12. Each tube should be used once only to avoid freezing and thawing as this can significantly decrease the virus titer. Use 1 mL aliquots tubes for virus amplification.
13. Other dilutions such as ½ log10 dilution could be used.
14. Change pipette tips between wells.
15. Avoid cell drying by minimizing the time between media aspiration and adding the virus inoculum or the serum-virus mixtures.
16. Alternatively, remove media from cells and fix cells with 100 μL ice-cold 4% paraformaldehyde for 5 min at room temperature. Remove fixative and stain cells with 100 μL crystal violet (0.05% w/v) in 20% methanol for 30 min at room temperature, and wash cells in tap water. Score wells as positive for MERS-CoV (i.e., no crystal violet) or negative for MERS-CoV (i.e., cells are stained with crystal violet).
17. For each serum sample, 12 μL are needed per single test; however, sera should be tested in at least duplicates, so more volume is needed. Different plates should be used when testing neutralization against different virus strains.
18. Set up back virus titration to ensure working virus concentration is accurate. Starting with the working virus dilution (2×10^3 $TCID_{50}$/mL), prepare twofold serial dilution in pre-warmed M-2 in a final volume of 60 μL (4 replicates per dilution). After dilution, add 60 μL of pre-warmed M-2 to each well for a final volume of 120 μL and incubate for 1 h in 37 °C incubator with 5% CO_2. Then, transfer 100 μL to Vero E6 cells in 96-well tissue culture plate and incubate for 3 days in 37 °C incubator with 5% CO_2. After incubation, examine the plate for CPE and calculate $TCID_{50}$ using Reed–Muench formula.

Acknowledgment

This work was supported by King Abdulaziz City for Science and Technology (KACST) through the MERS-CoV research grant program (number 09-1 to AMH), which is a part of the Targeted Research Program.

References

1. Zaki AM, van Boheemen S, Bestebroer TM et al (2012) Isolation of a novel coronavirus from a man with pneumonia in Saudi Arabia. N Engl J Med 367:1814–1820. https://doi.org/10.1056/NEJMoa1211721
2. Alagaili AN, Briese T, Mishra N et al (2014) Middle East respiratory syndrome coronavirus infection in dromedary camels in Saudi Arabia. MBio 5:e00884–e00814. https://doi.org/10.1128/mBio.00884-14
3. Azhar EI, El-Kafrawy SA, Farraj SA et al (2014) Evidence for camel-to-human transmission of MERS coronavirus. N Engl J Med 370:2499–2505. https://doi.org/10.1056/NEJMoa1401505
4. Poissy J, Goffard A, Parmentier-Decrucq E et al (2014) Kinetics and pattern of viral excretion in biological specimens of two MERS-CoV cases. J Clin Virol 61:275–278. https://doi.org/10.1016/j.jcv.2014.07.002
5. Müller MA, Meyer B, Corman VM et al (2015) Presence of Middle East respiratory syndrome coronavirus antibodies in Saudi Arabia: a nationwide, cross-sectional, serological study. Lancet Infect Dis 15:559–564. https://doi.org/10.1016/S1473-3099(15)70090-3
6. Corman VM, Albarrak AM, Omrani AS et al (2016) Viral shedding and antibody response in 37 patients with Middle East respiratory syndrome coronavirus infection. Clin Infect Dis 62:477–483. https://doi.org/10.1093/cid/civ951
7. Zhao J, Alshukairi AN, Baharoon SA et al (2017) Recovery from the Middle East respiratory syndrome is associated with antibody and T-cell responses. Sci Immunol 2:eaan5393. https://doi.org/10.1126/sciimmunol.aan5393
8. Al-Amri SS, Abbas AT, Siddiq LA et al (2017) Immunogenicity of candidate MERS-CoV DNA vaccines based on the spike protein. Sci Rep 7:44875. https://doi.org/10.1038/srep44875
9. Perera RA, Wang P, Gomaa MR et al (2013) Seroepidemiology for MERS coronavirus using microneutralisation and pseudoparticle virus neutralisation assays reveal a high prevalence of antibody in dromedary camels in Egypt, June 2013. Euro Surveill 18:20574
10. Reusken C, Mou H, Godeke GJ et al (2013) Specific serology for emerging human coronaviruses by protein microarray. Euro Surveill 18:20441
11. Al-Abdallat MM, Payne DC, Alqasrawi S et al (2014) Hospital-associated outbreak of Middle East respiratory syndrome coronavirus: a serologic, epidemiologic, and clinical description. Clin Infect Dis 59:1225–1233. https://doi.org/10.1093/cid/ciu359
12. Hemida MG, Perera RA, Al Jassim RA et al (2014) Seroepidemiology of Middle East respiratory syndrome (MERS) coronavirus in Saudi Arabia (1993) and Australia (2014) and characterisation of assay specificity. Euro Surveill 19:20828
13. Kleine-Weber H, Elzayat MT, Wang L et al (2015) Inability of rat DPP4 to allow MERS-CoV infection revealed by using a VSV pseudotype bearing truncated MERS-CoV spike protein. Arch Virol 160:2293–2300. https://doi.org/10.1128/JVI.01381-18
14. Park SW, Perera RA, Choe PG et al (2015) Comparison of serological assays in human Middle East respiratory syndrome (MERS)-coronavirus infection. Euro Surveill 20:30042. https://doi.org/10.2807/1560-7917.ES.2015.20.41.30042
15. Fukushi S, Fukuma A, Kurosu T et al (2017) Characterization of novel monoclonal antibodies against the MERS-coronavirus spike protein and their application in species-independent antibody detection by competitive ELISA. J Virol Methods 251:22–29. https://doi.org/10.1016/j.jviromet.2017.10.008
16. Trivedi S, Miao C, Al-Abdallat MM et al (2017) Inclusion of MERS-spike protein ELISA in algorithm to determine serologic evidence of MERS-CoV infection. J Med Virol 90:367–371. https://doi.org/10.1002/jmv.24948
17. Hashem AM, Al-Amri SS, Al-Subhi TL et al (2019) Development and validation of different indirect ELISAs for MERS-CoV serological testing. J Immunol Methods 466:41–46. https://doi.org/10.1016/j.jim.2019.01.005
18. Reed LJ, Muench H (1938) A simple method of estimating fifty per cent endpoints. Am J Epidemiol 27:493–497

Chapter 10

Generation of MERS-CoV Pseudotyped Viral Particles for the Evaluation of Neutralizing Antibodies in Mammalian Sera

Abdulrahman Almasaud, Naif Khalaf Alharbi, and Anwar M. Hashem

Abstract

Pseudotyped viral particle production has been used extensively and broadly for many viruses to evaluate levels of neutralizing antibodies, viral entry inhibitors and vaccine immunogenicity. This assay is extremely safe and useful alternative to live virus-based assay without the need for high containment facilities. In this chapter, we describe the generation of MERS-CoV pseudotyped viral particles (MERSpp) expressing full-length spike protein using second-generation lentiviral packaging system. This platform is optimized to generate high titer of MERSpp and to test sera from different mammalian species.

Key words MERS-CoV, Pseudotype viral particles, Pseudovirus, Transfection, Neutralization assay

1 Introduction

Pseudotype viruses (also known as pseudoviruses, pseudoparticles, or pseudotype viral particles) were discovered in 1911 [1]. They are chimeric viral particles expressing recombinant glycoproteins from one virus on the surface of another replication-deficient virus (viral vector), generating a single round chimeric viral particles. Pseudotype viral particles have been developed for many viruses especially those requiring high containment facilities such as SARS-CoV, MERS-CoV, Ebola, and highly pathogenic influenza A viruses without the need to handle wild-type viruses [2]. Pseudotype-based assays allow for accurate, specific, and sensitive detection of neutralizing antibodies (nAbs) and screening for viral entry inhibitors.

The most efficient and common viral vectors for the production of pseudotype viral particles are lentiviruses. Lentiviral vectors are retroviruses, which are enveloped single-stranded RNA viruses, derived, for example, from human immunodeficiency virus type 1 (HIV-1). They have been used to develop pseudotype viral particles for many pathogenic viruses [3–6]. These replication-

Rahul Vijay (ed.), *MERS Coronavirus: Methods and Protocols*, Methods in Molecular Biology, vol. 2099, https://doi.org/10.1007/978-1-0716-0211-9_10,

deficient vectors offer a number of advantages including that they do not replicate in mammalian cells, they infect dividing and non-dividing cells, they can incorporate large transgenes derived from other pathogenic viruses as large as 9 kb, and they induce no or weak immune response [7–9]. Several studies have utilized lentiviral vectors to generate MERS-CoV pseudotype viral particles (MERSpp) to evaluate nAbs in humans and animals. Here, we describe detailed protocol for the generation and utilization of MERSpp. This protocol is based on lentiviral system and use of luciferase enzyme as the main readout reporter of the system.

2 Materials

2.1 General Materials

1. Biosafety cabinet.
2. Inverted microscope.
3. Low-speed centrifuge.
4. 37 °C incubator with 5% CO_2.
5. 37 °C water bath.
6. Sterile tissue culture 75 cm^2 flasks.
7. Sterile serological pipettes.
8. Sterile 15 mL falcon tubes.
9. Pipettes.
10. Multichannel pipette.
11. Sterile disposable aerosol-resistant filtered tips.
12. Complete cell growth media: Filter-sterilized Dulbecco's modification of Eagle medium (DMEM) supplemented with 10% heat-inactivated fetal bovine serum (FBS), 1% L-glutamine, and 1% penicillin/streptomycin.
13. 70% Ethanol for decontamination of laminar flow biosafety cabinet and objects brought into the hood.
14. Sterile DPBS without calcium or magnesium.
15. Sterile 1× trypsin-EDTA in DPBS without calcium or magnesium.

2.2 Making MERSpp

1. HEK 293T cells.
2. Envelope DNA plasmid: Glycoprotein expression plasmid: pCAGGS-MERS-CoV spike (*see* **Note 1**).
3. Lentiviral plasmid expressing firefly luciferase: pCSFLW [9] (*see* **Note 1**).
4. Lentiviral packaging plasmid: second-generation packaging vector expressing HIV-Gag-Pol: p8.91 [10] (*see* **Note 1**).
5. Branched polyethylenimine solution (PEI) (1 mg/mL) (*see* **Note 2**).

6. Opti-MEM reduced serum media.
7. 1 M HEPES.
8. Sterile 1.5 mL microcentrifuge tubes (two tubes per transfection).
9. Sterile 0.22 μm, γ-irradiated syringe filters.
10. Sterile 0.45 μm γ-irradiated syringe filters.
11. Sterile 10 mL syringes.

2.3 Titration Step

1. 96-Well white opaque culture plate: Sterile luminomer IsoPlate-96 B & W tissue culture plate with lid.
2. Supernatant containing MERSpp pseudotype virus, VSV-G or ΔEnv pseudotype viruses.
3. Confluent Huh7 cells in a tissue culture 75 cm^2 flask, cultured in complete cell growth media (preferentially subcultured at 1:4 ratio 48 h before use).
4. Cell Counter or hemocytometer.
5. Bright Glo™ kits (Promega).
6. Luminometer.

2.4 Neutralization Step

1. 96-Well white opaque culture plate: Sterile luminomer IsoPlate-96 B & W tissue culture plate with lid.
2. Supernatant containing titrated pseudotyped viruses.
3. Mammalian serum samples to be tested.
4. 5 mL Huh7 cells suspension at 2×10^5cells/mL in complete cell growth media.
5. Bright Glo™ kits (Promega).
6. Luminometer.

3 Methods

Proper antiseptic techniques should be used, and all equipment and solutions to be used with cells must be sterile. All steps should be performed under biosafety cabinet class II in a tissue culture room. All cell culture incubations should be performed in a humidified 37 °C incubator with 5% CO_2, and all solutions should be pre-warmed to room temperature or 37 °C before use with cells.

3.1 Generation of Pseudotype Viruses

1. Plate 293T cells 24 h before transfection in a 75 cm^2 tissue culture flask and incubate the flask in 37 °C incubator with 5% CO_2 to be 70% confluent at the time of transfection (next day) (*see* **Note 3**).
2. On the day of transfection, prepare and label two sterile 1.5 mL microcentrifuge tubes (tube 1 and tube 2) per transfection.

3. Add 200 μL pre-warmed Opti-MEM to tube 1.
4. Add the DNA plasmids to tube 1 at the ratio 0.9:1:1.5 (pCAGGS-MERS-CoV spike:p8.91:pCSFLW) (*see* **Note 4**).
5. Add 200 μL Opti-MEM and 35 μL of 1 mg/mL PEI to tube 2 (*see* **Note 5**).
6. Mix both tubes by gentle mixing and incubate for 5 min at room temperature.
7. Transfer the content of tube 2 into tube 1 and incubate the tube at room temperature for 20 min. Mix the tube by gentle rocking every 3–4 min during incubation period.
8. During incubation, remove the media from the 293 T cell flask and add 7 mL of fresh pre-warmed complete cell growth media (*see* **Note 6**).
9. After the 20 min incubation, pipette the mixture from the tube onto 293T cells dropwise over the complete area of the flask. Swirl the flask gently to ensure even dispersal.
10. Incubate the flask at 37 °C, 5% CO_2 for overnight for 12–16 h.
11. After incubation, change the media by removing the old media and adding 7 mL of fresh pre-warmed complete cell growth media (*see* **Note 6**).
12. Incubate the flask at 37 °C, 5% CO_2 for additional 32–36 h.
13. Collect the supernatant, which contains the viral pseudotype particles, using sterile 10 mL sterile syringe.
14. Filter the collected supernatant through a Sterile 0.45 μm filter into a sterile 15 mL tube.
15. Store the filtered supernatant at −80 °C (*see* **Note 7**).

3.2 MERSpp Titration

1. In a 96-well white opaque culture plate, add 50 μL of pre-warmed complete cell growth media to all wells in column 12 "cell only control" (CC) as a negative control.
2. Add 50 μL of pre-warmed complete cell growth media to all wells in rows B to H, columns 1 to 11 (Fig. 1).
3. Add 100 μL of supernatant containing MERSpp, VSV-G, or ΔEnv pseudotype viruses to wells of row A as shown in Fig. 1 excluding CC column (i.e., column 12) (*see* **Note 8**).
4. Remove 50 μL from virus-containing wells in row A (A1-A11) and perform 1:2 serial dilutions downward to all wells below (Fig. 1).
5. During each dilution step mix well by pipetting eight times up and down.
6. Continue the dilution until row H and discard the final 50 μL from the last wells in row H (Fig. 1) (*see* **Note 9**).
7. Harvest Huh7 cells from the 75 cm^2 tissue culture flask using standard trypsinization procedure (*see* **Note 10**).

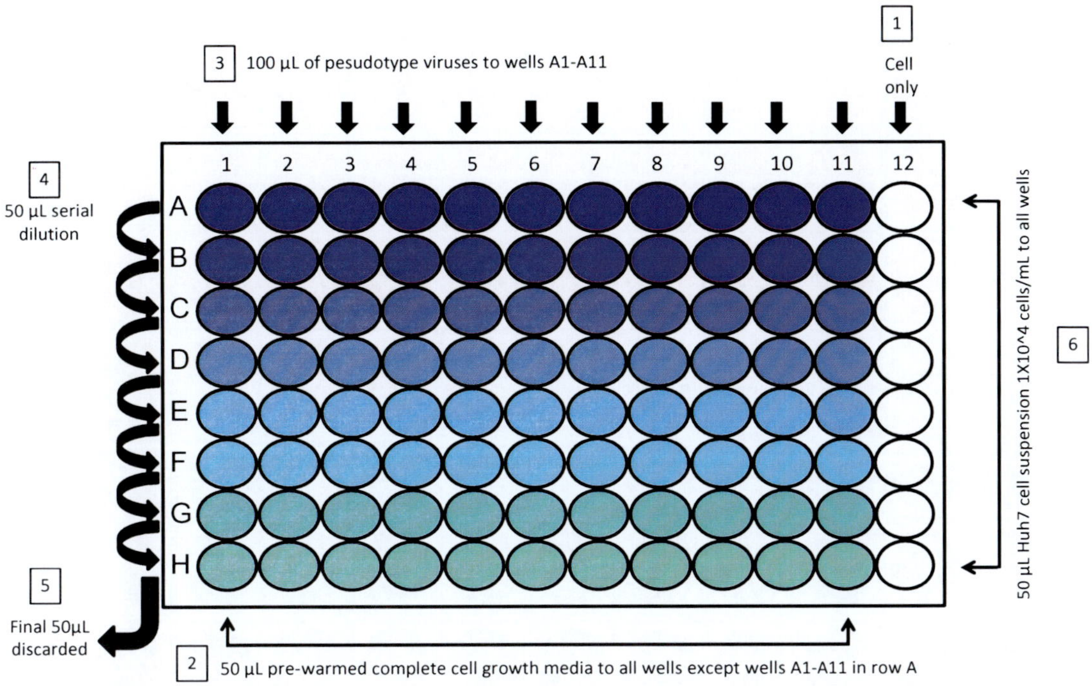

Fig. 1 Plate layout for MERSpp titration. The sequential steps of plate preparation are indicated by numbered boxes. The color intensity indicates the expected values of luciferase readout

8. Count the cells and prepare a 5 mL cell suspension at 2×10^5cells/mL in pre-warmed complete cell growth media (i.e., every 50 μL should contain 1×10^4cells in total).
9. Add 50 μL of the Huh7 cell suspension to all well in the 96-well white opaque culture plate (*see* **Note 11**).
10. Centrifuge the 96-well opaque culture plate for 1 min at $500 \times g$ to settle down any droplets on the inner sides of the wells.
11. Incubate the plate for 48 h at 37 °C, 5% CO_2 (*see* **Note 12**).
12. In 15 mL falcon tube, prepare 1:1 Bright Glo™ luciferase substrate by adding equal amount of the substrate and pre-warmed complete cell growth media (for one 96-well plate: add 2.5 mL of substrate and 2.5 mL pre-warmed complete cell growth media to make 5 mL total volume of prepared substrate).
13. After incubation, pipette out and discard supernatant from all wells and add 50 μL of the prepared substrate into each well of the 96-well plate (*see* **Notes 13** and **14**).
14. Wait 5 min and then read the plate using a luminometer and save the results. Figure 2 shows example readout of a titration plate (*see* **Note 15**).

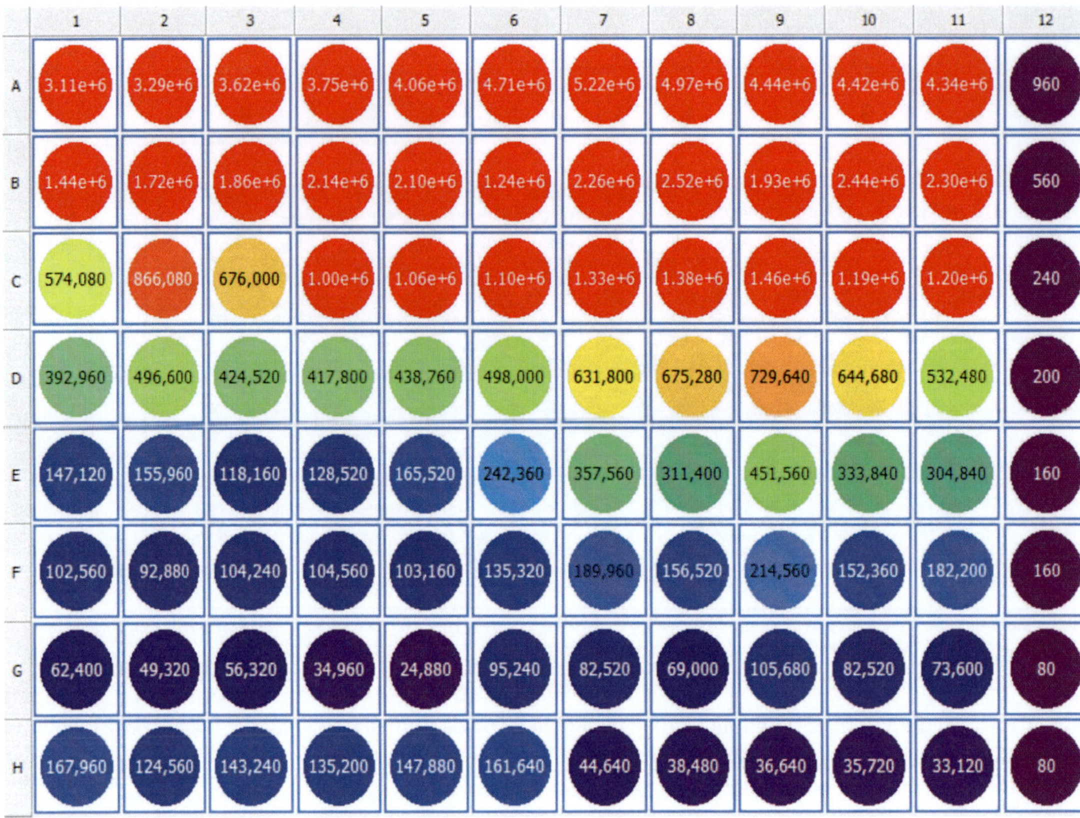

Fig. 2 An example readout of MERSpp titration plate

3.3 MERSpp Neutralization (MN) Assay

1. Add 100 μL of pre-warmed complete cell growth media to all wells in column 12 "cell only control" (CC) of the 96-well opaque white plate (Fig. 3).
2. Add 50 μL of pre-warmed complete cell growth media to all wells in column 11 "cells + virus control" (VC) (Fig. 3).
3. Add 50 μL of pre-warmed complete cell growth media to all wells in rows B to H in columns 1 to 10 (Fig. 3).
4. Add 95 μL of pre-warmed complete cell growth media in wells in row A (A1–A10) (Fig. 3).
5. Add 5 μL of serum samples in duplicate (two wells per sample) in wells in row A (A1–A10) (Fig. 3) to have 1:20 dilution (*see* **Note 16**).
6. Remove 25 μL from wells in row A (A1–A10) and perform 1:3 serial dilutions downward to all wells below (Fig. 3).
7. During each dilution step mix well by pipetting five times up and down.

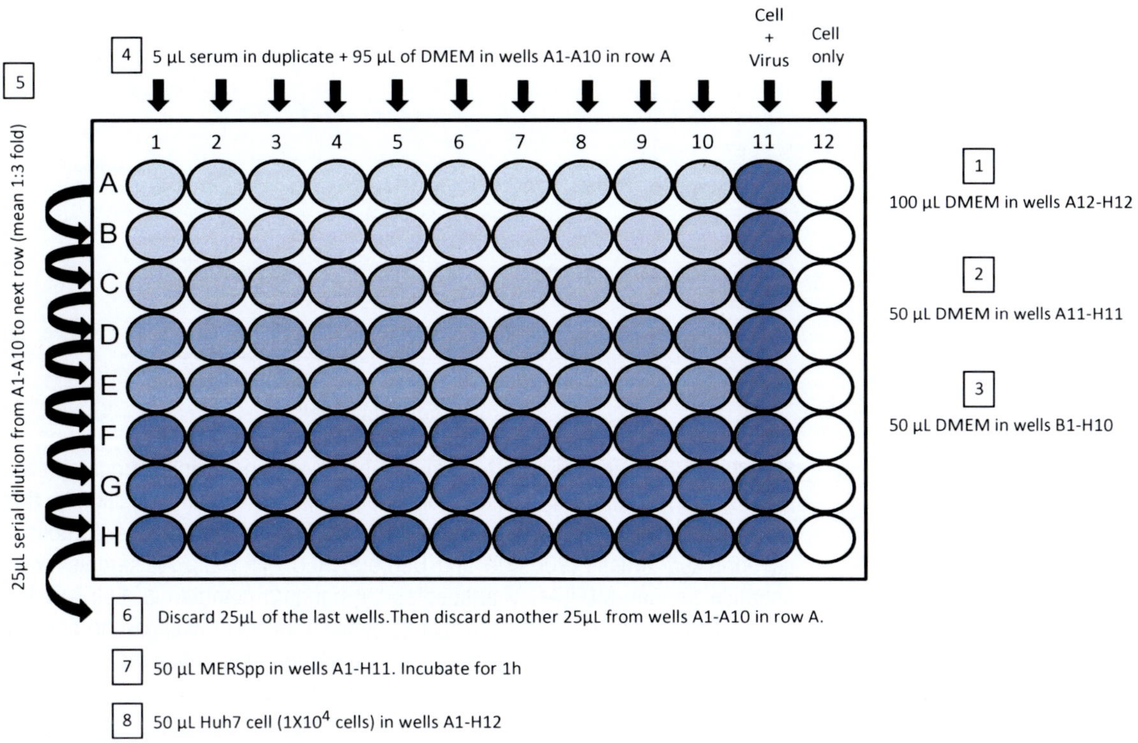

Fig. 3 Plate preparation for MERSpp neutralization assay. The preparation steps for MERSpp neutralization assay plate are indicated by numbered boxes. The color intensity indicates the expected values of luciferase readout

8. Continue the dilution until row H; discard the final 25 μL from the last wells in row H (Fig. 3).
9. Also discard 25 μL from wells in row A (A1–A10).
10. Based on MERSpp titration, prepare a MERSpp suspension with a concentration of 200,000 RLU per 50 μL, a total of 5 mL are needed for one 96-well plate (*see* **Note 15**).
11. Add 50 μL of MERSpp suspension into each well in the plate except column 12 (CC).
12. Incubate the plate for 1 h at 37 °C, 5% CO_2.
13. After incubation, add 50 μL of the Huh7 cell suspension to all wells (1×10^4cells in total). So, each well in the plate will have 150 μL total volume.
14. Incubate the plate for 48 h at 37 °C, 5% CO_2 (*see* **Note 12**).
15. After incubation, discard the supernatant from all wells and measure luciferase activity on a luminometer as indicated in the titration step and save the results (*see* **Note 13**). Figure 4 shows an example of the levels of background neutralization activity from different species.

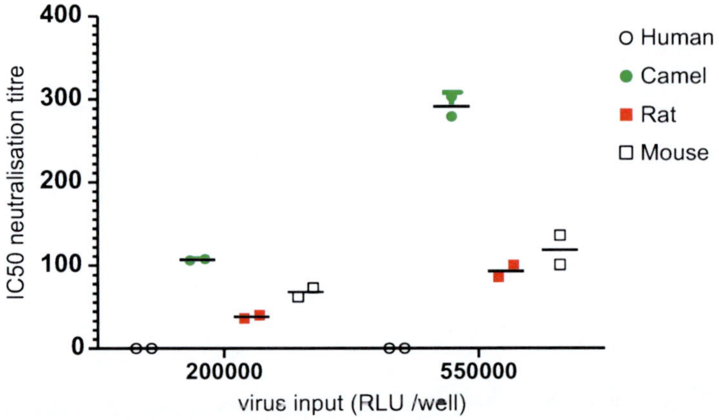

Fig. 4 MERSpp neutralization assay for naïve serum samples from different species. Serum samples from different species, including humans, camels, rats, and mice, were tested in a MERSpp NA with two different concentrations of pseudotyped viruses (200,000 and 550,000 RLU per well). All samples were negative for Anti-MERS-CoV antibodies by standard commercial ELISA before conducting the MERSpp NA. The results showed levels of background neutralization that varied between species. This background neutralization is expected based on our previous observation (unpublished data)

4 Notes

1. Other mammalian vectors such as pcDNA3.1 could be used. Use codon-optimized transgene for mammalian cell expression. All plasmids need to be transformed into DH5α cells or similar cells using ampicillin as a selection antibiotic. Plasmids can be purified using routine protocols.
2. To prepare branched polyethylenimine (PEI) at 1 mg/mL, dissolve PEI in endotoxin-free water (pre-warmed to 80 °C). Let it cool down at room temperature, and then neutralize the pH (pH 7.0) using 1 M HEPES buffer to a final concentration of 15 mM. Sterilize the solution by filtration using Sterile 0.22 μm filters. Filtration is important not only for sterility but also to remove undissolved PEI, which could reduce the transfection efficiency. Store aliquots of 0.2–1 mL at −80 °C for long-term storage. Thawed and working solutions could be stored at 4 °C for up to 2 months.
3. Alternatively, 100 mm petri dishes could be used. Cells could be cultured without antibiotics to reduce toxicity and cell death.
4. A plasmid encoding vesicular stomatitis virus G protein (VSV-G-pcDNA3.1) or an empty pcDNA3.1 vector could be used instead of pCAGGS-MERS-CoV spike to generate VSV-G pseudovirus or pseudovirus without Env (ΔEnv) as controls.

5. Alternatively, Lipofectamine 2000 could be used. To tube 2, add 15 μL of Lipofectamine 2000 into 200 μL pre-warmed Opti-MEM. It is recommended to use a range of 0.5–5 μL of Lipofectamine 2000 per μg of DNA.
6. Make sure to add media slowly to one side of the flask to avoid detaching adherent cells. Avoid cell drying by adding the media right after removing the old media.
7. It is recommended to aliquot the collected supernatant to avoid multiple freezing and thawing. Collected supernatant could be stored at 4 °C for up to one week without loss in MERSpp titer.
8. Use of VSV-G or ΔEnv pseudotype viruses is optional as controls.
9. After completing the serial dilutions, the final volume per well should be 50 μL of mixed cell growth media and supernatant containing viral pseudotype particles.
10. To harvest or maintain Huh7 cells, remove media from the flask, wash the cell monolayer gently with 3–5 mL of sterile pre-warmed DPBS without calcium or magnesium, and discard the used washing solution. Add 3 mL pre-warmed 1× trypsin-EDTA in DPBS without calcium or magnesium to the cell monolayer and incubate for 5 min at 37 °C, 5% CO_2 to detach cells (incubation may vary, so check the cells every 2–3 min). After cells are detached, add 3 mL pre-warmed complete cell growth media to the flask to quench trypsin activity, and collect detached cells in 15 mL sterile falcon tube. Make sure to centrifuge the collected cells and discard the supernatant. Then, add new 6 mL pre-warmed complete cell growth media and re-suspend the cells by pipetting up and down to make a homogenous cell suspension.
11. It is recommended to seed a few wells of a regular 96-well tissue culture plate with 50 μL of the Huh7 cell suspension and 50 μL pre-warmed complete cell growth media to be able to check cells under the microscope to get a sense of how confluent and viable the cells are.
12. Ensure the incubator is H_2O saturated to avoid cell drying.
13. Make sure to avoid detaching and removing adherent cells when aspirating off medium from wells. Therefore, it is recommended to only remove 90 μL and 140 μL from each well in the titration and the neutralization plates, respectively.
14. Alternatively, aspirate off medium and add 30 μL of 1× passive lysis buffer (25 mM Tris-phosphate (pH 7.8), 2 mM DTT, 2 mM 1,2-diaminocyclohexaneN,N,N′,N′-tetraacetic acid, 10% glycerol, 1% Triton X-100) to each well. Then, freeze and thaw the cells once on dry ice. After thawing, leave the plate at room temperature for 30 min before reading luciferase

activity on Luminometer supplied with two injectors for both luciferase buffer (15 mM MgSO4, 15 mM KPO4 (pH 7.8), 1 mM ATP, 1 mM DTT in d H_2O; 20 mL is needed per plate to add 100 μL per well) and luciferase substrates (D-luciferin potassium salt dissolve in H_2O at 0.3 mg/mL, 10 mL is needed per plate to add 50 μL per well). Measure light produced from the reaction ~8 s after adding the substrate using an integration time of 5–30 s.

15. Alternatively, the HIV-1 p24 content in the generated pseudotyped viruses could be quantified by HIV-1 $p24^{CA}$ capture ELISA kit and pseudotyped viruses equivalent to ~500 ηg of p24 could be used in neutralization assay.
16. Use heat-inactivated serum samples at 56 °C for 30 min.

Acknowledgment

We would like to express our gratitude and thanks to Dr. Nigel Temperton and Dr. Keith Grehan for generating and providing the required plasmids for MERSpp. This work was supported by King Abdulaziz City for Science and Technology (KACST) through the MERS-CoV research grant program (number 09-1 and 34-01 to AMH and NKA, respectively), which is a part of the Targeted Research Program.

References

1. Rous P (1910) A transmissible avian neoplasm. (Sarcoma of the common fowl). J Exp Med 12:696–705
2. Carnell GW, Ferrara F, Grehan K et al (2015) Pseudotype-based neutralization assays for influenza: a systematic analysis. Front Immunol 6:16. https://doi.org/10.3389/fimmu.2015.00161
3. Ferrara F, Temperton N (2017) Chimeric influenza haemagglutinins: generation and use in pseudotype neutralization assays. Methods 4:11–24. https://doi.org/10.1016/j.mex.2016.12.001
4. Urbanowicz RA, McClure CP, King B et al (2016) Novel functional hepatitis C virus glycoprotein isolates identified using an optimized viral pseudotype entry assay. J Gen Virol 97:2265–2279. https://doi.org/10.1099/jgv.0.000537
5. Chin AWH, Perera RAPM, Guan Y et al (2015) Pseudoparticle neutralization assay for detecting Ebola-neutralizing antibodies in biosafety level 2 settings. Clin Chem 61:885–886. https://doi.org/10.1373/clinchem.2015.238204
6. Grehan K, Ferrara F, Temperton N (2015) An optimised method for the production of MERS-CoV spike expressing viral pseudotypes. Methods 2:379–384. https://doi.org/10.1016/j.mex.2015.09.003
7. Cronin J, Zhang X-Y, Reiser J (2005) Altering the tropism of lentiviral vectors through pseudotyping. Curr Gene Ther 5:387–398
8. Schnierle BS, Stitz J, Bosch V et al (1997) Pseudotyping of murine leukemia virus with the envelope glycoproteins of HIV generates a retroviral vector with specificity of infection for CD4-expressing cells. Proc Natl Acad Sci U S A 94:8640–8645
9. Zufferey R, Dull T, Mandel RJ et al (1998) Self-inactivating lentivirus vector for safe and efficient in vivo gene delivery. J Virol 72:9873–9880
10. Unseld M, Marienfeld JR, Brandt P et al (1997) The mitochondrial genome of Arabidopsis thaliana contains 57 genes in 366,924 nucleotides. Nat Genet 15:57–61

Chapter 11

Qualitative and Quantitative Determination of MERS-CoV S1-Specific Antibodies Using ELISA

Sawsan S. Al-amri and Anwar M. Hashem

Abstract

Indirect enzyme-linked immunosorbent assay (ELISA) enables detection and quantification of antigen-specific antibodies in biological samples such as human or animal sera. Most current MERS-CoV serological assays such as neutralization, immunofluorescence, or protein microarray rely on handling of live MERS-CoV in high containment laboratories, highly trained personnel as well as the need for expensive and special equipment and reagents representing a hurdle for most laboratories especially when resources are limited. In this chapter, we describe a validated and optimized indirect ELISA protocol based on recombinant S1 subunit (amino acids 1–725) of MERS-CoV for qualitative and quantitative determination of MERS-CoV-binding antibodies.

Key words Antigens, Antibodies, Serology, ELISA, MERS-CoV, Recombinant S1 subunit

1 Introduction

Enzyme-linked immunosorbent assay (ELISA) is an immunological technique used to study the interaction between antigens and antibodies. The use of an enzyme-linked conjugate as well as substrate leads to changes in color which help to determine the presence and the quantity of substances such as peptides, proteins, antibodies, and hormones in a given sample, such as blood or urine samples [1]. The importance of validated ELISA is evident especially when dealing with pathogens requiring high containment facilities such as MERS-CoV as it could provide a rapid, simple, and cheap method for field or clinical use without the need for high containment laboratories.

Different indirect ELISAs based on MERS-CoV nucleocapsid (N) or spike (S) proteins were developed and used in epidemiological and surveillance studies [2–8]. Here, we provide a detailed protocol for indirect ELISA based on recombinant MERS-CoV S1 subunit (amino acids 1–725) for qualitative and quantitative MERS-CoV serological testing. This assay was developed and

Rahul Vijay (ed.), *MERS Coronavirus: Methods and Protocols*, Methods in Molecular Biology, vol. 2099,
https://doi.org/10.1007/978-1-0716-0211-9_11,

validated using a large number of well-characterized human serum samples [8] and could be adapted by any laboratory especially that all required reagents are commercially available.

2 Materials

Prepare all solutions using deionized water and store them at room temperature or 4 °C. Diligently follow all waste disposal regulations when disposing waste materials.

2.1 General Materials

1. 1 L graduated cylinder.
2. 100 mL graduated cylinder.
3. Weighing balance.
4. Magnetic stirrer plate.
5. Magnetic stirrers.
6. PH meter.
7. 1 L glass bottles.
8. 15 and 50 mL falcon tubes.
9. Sterile U-shaped 96-well plate.
10. Multichannel pipette.
11. Sterile disposable tips.
12. Reagent reservoirs (*see* **Note 1**).
13. MaxiSorp flat-bottom 96-well ELISA plates.
14. ELISA plate sealing covers.
15. Automated 96-well plate washer.
16. Automated microplate reader.

2.2 Solution Preparation

1. Phosphate-buffered saline (PBS; 10×): Weigh and transfer 80 g NaCl, 2 g KCl, 14.4 g Na_2HPO_4, and 2.4 g KH_2PO_4 to a 1 L graduated cylinder, and add about 800 mL water. Mix for about 15 min and adjust pH to 7.4 with HCl. Make up to 1 L with water. Autoclave the solution and store at room temperature.
2. PBS; 1×: Add 100 mL of 10× PBS to a 1 L graduated cylinder and complete the volume to 1 L by adding 900 mL water. Transfer the buffer to a 1 L glass bottle and store at 4 °C.
3. Coating buffer: PBS; 1× (*see* **Note 2**).
4. Wash buffer: 1× PBS containing 0.1% Tween-20 (TBST). Add 1.0 mL Tween-20 to 1 L 1× PBS (0.1% v/v). Mix and store at 4 °C.
5. Blocking buffer: 5% skim milk in PBST (*see* **Note 3**). Store at 4 °C.

6. Diluent for primary and secondary antibodies: Primary and secondary antibodies should be diluted in blocking buffer (*see* **Note 3**). Store at 4 °C.

2.3 Antigen and Conjugates

1. Recombinant MERS-CoV S1 subunit protein.
2. Primary antibody (human or animal serum samples).
3. Secondary antibody (horseradish peroxidase (HRP)-conjugated anti-human IgG antibody could be used if testing human sera).
4. KPL SureBlue tetramethylbenzidine (TMB) substrate.
5. KPL TMB BlueSTOP solution.

3 Methods

Carry out all procedures at room temperature unless otherwise specified.

3.1 Qualitative Detection of MERS-CoV S1-Specific Antibodies

1. Dilute recombinant MERS-CoV S1 subunit protein (antigen) to a final concentration of 1 μg/mL using 1× PBS buffer. You need 10 mL per 96-well ELISA plate.
2. Transfer the diluted recombinant protein to a clean reagent reservoir and coat the 96-well ELISA plate with 100 μL/well using multichannel pipette.
3. Seal the plate using ELISA plate sealing cover (adhesive film) and incubate at 4 °C overnight to allow the antigen to adsorb to the well surface (*see* **Note 4**).
4. Wash the 96-well ELISA plate three times using automated 96-well plate washer with PBST. Use 350 μL wash buffer per well (*see* **Note 5**).
5. After washing, invert the plate and tap it firmly on an absorbent paper to remove any residual liquid.
6. Block the plate with blocking buffer (300 μL/well), seal the plate and incubate for 1 h at room temperature (*see* **Note 6**).
7. During incubation, dilute serum samples in blocking buffer in duplicate at a final dilution of 1:200 or 1:400 (*see* **Notes** 7 and **8**).
8. Wash the plate as indicated in **steps 4** and **5**.
9. Add 100 μL of diluted serum samples or controls to each well, seal the plate, and incubate at 37 °C for 1 h (*see* **Note 9**).
10. During incubation, dilute appropriate HRP-conjugated secondary antibody in blocking solution according to manufacturer's instructions.
11. Wash the plate as indicated in **steps 4** and **5** (*see* **Note 10**).

12. Add 100 μL of diluted secondary antibody to each well and incubate at 37 °C for 1 h.
13. Wash the plate as indicated in **steps 4** and **5** (*see* **Note 11**).
14. Add 100 μL of TMB substrate to each well and incubate at room temperature for 30 min in the dark for colorimetric development (*see* **Notes 12** and **13**).
15. Stop reaction with equal volume of TMB BlueSTOP solution.
16. Read absorbance in the plate on automated microplate reader at 630 nm (*see* **Note 14**).
17. Samples with absorbance above the cutoff value of 0.34 are considered positive (assay sensitivity and specificity are 94.9% and 95.2%, respectively [8]) (*see* **Note 15**).

3.2 End-Point Titration of MERS-CoV S1-Specific Antibodies

1. Coat and block the plate as indicated in the **steps 1–6** in Subheading 3.1.
2. In a new sterile U-shaped 96-well plate, add 297 μL blocking buffer to all wells in column 1 (*see* **Note 16**).
3. Add 150 μL blocking buffer to all remaining wells in the plate (Fig. 1).
4. Add 3 μL from each serum per well in all wells in column 1 to have 1:100 dilution (Fig. 1). Test each serum sample in duplicates (*see* **Notes** 7 and **8**).

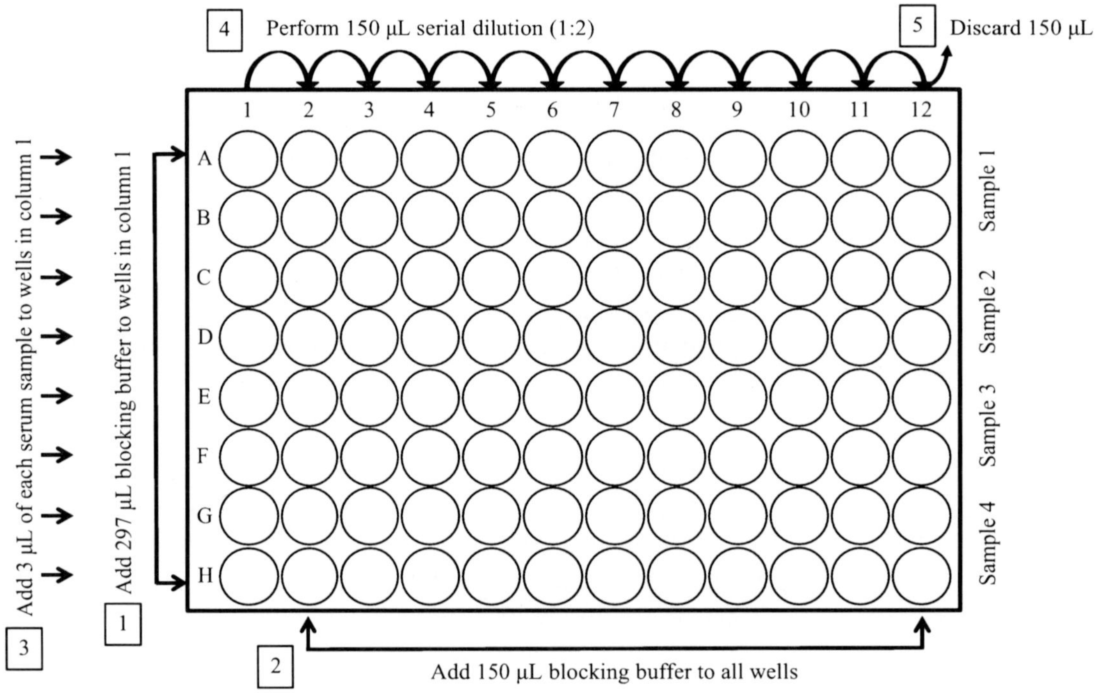

Fig. 1 Plate layout for serum sample serial dilution

5. Perform twofold serial dilutions by transferring 150 μL progressively from column to column using a multichannel pipette (Fig. 1).
6. During each dilution step mix well by pipetting eight times up and down (*see* **Note 17**).
7. Discard the final 150 μL after column 12.
8. Wash the incubated plate as indicated in **steps 4** and **5** in Subheading 3.1 to remove blocking buffer.
9. Add 100 μL from each dilution to each well using a multichannel pipette, seal the plate, and incubate at 37 °C for 1 h (*see* **Note 17**).
10. Continue using **steps 10–16** from Subheading 3.1 and save the results.
11. The last dilution needs to reach a signal equivalent to the background reading from blocking buffer without serum (*see* **Note 18**).
12. The every last dilution that gives twice the signal of the background indicates the end-point titer of the sample; otherwise, antibody titer can be determined using four-parameter logistic (4PL) regression curve in SigmaPlot or GraphPad Prism software.

4 Notes

1. Use different reagent reservoir for each buffer.
2. Alternatively, carbonate-bicarbonate buffer (50 mM), pH 9.6, could be used as coating buffer: Prepare buffer by adding 2.88 g of sodium bicarbonate ($NaHCO_3$) and 1.67 g of sodium carbonate (Na2CO3) to ~980 mL water. Adjust the pH to 9.6 with HCl if needed and complete the volume to 1 L with water.
3. Measure 100 mL of 1× PBST to a 100 mL graduated cylinder and transfer the volume to a glass bottle. Transfer 5 g skim milk powder into the bottle and stir until dissolved.
4. Coating buffer helps to bind antigen to the wells. During coating, sealing the plates will help prevent any reagents from evaporating overnight when leaving them in the refrigerator.
5. Washing the 96-well ELISA plate with PBST will help remove any unbound antigens from the wells.
6. Blocking helps in preventing nonspecific binding of detection antibodies to the microplate surface, reducing signal background and improving the signal-to-noise ratio. Blocking could be done at 4 °C for overnight.

7. Use heat-inactivated serum samples at 56 °C for 30 min.
8. Mix thawed serum samples before and after dilution with a vortex for about 10 s.
9. No primary antibody control could be included by adding 100 μL of blocking buffer per well.
10. This washing step will help removing nonspecific or unbound antibodies.
11. This washing step is critical to reduce background signal.
12. Warm TMB substrate and stop solution to room temperature before use. Never pipette directly from the bottle. Pour out needed amount into a plastic reservoir. Do not return excess to the primary container.
13. Avoid shaking.
14. Stopped reactions should be read within 30 min. TMB Blue-STOP allows the chromophore to remain blue, instead of turning yellow. If using a stop solution resulting in a yellow reaction, read the plate at 450 nm.
15. Samples with absorbance values that fall 0.26 and 0.34 should be considered "indeterminate" and should be validated with other methods if possible.
16. Do **steps 2**–7 during incubation with blocking buffer.
17. Change pipette tips between wells.
18. Higher dilutions of the samples may be required in case the last dilution did not reach a signal equivalent to the background.

Acknowledgment

This work was supported by King Abdulaziz City for Science and Technology (KACST) through the MERS-CoV research grant program (number 09-1 to AMH), which is a part of the Targeted Research Program.

References

1. Aydin S (2015) A short history, principles, and types of ELISA, and our laboratory experience with peptide/protein analyses using ELISA. Peptides 72:4–15. https://doi.org/10.1016/j.peptides.2015.04.012
2. Al-Abdallat MM, Payne DC, Alqasrawi S et al (2014) Hospital-associated outbreak of Middle East respiratory syndrome coronavirus: a serologic, epidemiologic, and clinical description. Clin Infect Dis 59:1225–1233. https://doi.org/10.1093/cid/ciu359
3. Müller MA, Meyer B, Corman VM et al (2015) Presence of Middle East respiratory syndrome coronavirus antibodies in Saudi Arabia: a nationwide, cross-sectional, serological study. Lancet Infect Dis 15:559–564. https://doi.org/10.1016/S1473-3099(15)70090-3
4. Park SW, Perera RAPM, Choe PG et al (2015) Comparison of serological assays in human Middle East respiratory syndrome (MERS)-coronavirus infection. Euro Surveill 20

(41):30042. https://doi.org/10.2807/1560-7917.ES.2015.20.41.30042

5. Al-Amri SS, Abbas AT, Siddiq LA et al (2017) Immunogenicity of candidate MERS-CoV DNA vaccines based on the spike protein. Sci Rep 7:44875. https://doi.org/10.1038/srep44875
6. Chan JF, Sridhar S, Yip CC et al (2017) The role of laboratory diagnostics in emerging viral infections: the example of the Middle East respiratory syndrome epidemic. J Microbiol 55:172–182. https://doi.org/10.1007/s12275-017-7026-y
7. Trivedi S, Miao C, Al-Abdallat MM et al (2017) Inclusion of MERS-spike protein ELISA in algorithm to determine serologic evidence of MERS-CoV infection. J Med Virol 90:367–371. https://doi.org/10.1002/jmv.24948
8. Hashem AM, Al-amri SS, Al-subhi TL et al (2019) Development and validation of different indirect ELISAs for MERS-CoV serological testing. J Immunol Methods 466:41–46. https://doi.org/10.1016/j.jim.2019.01.005

PART IV

Mouse Models for MERS–Coronavirus

Chapter 12

Genetically Engineering a Susceptible Mouse Model for MERS-CoV-Induced Acute Respiratory Distress Syndrome

Sarah R. Leist and Adam S. Cockrell

Abstract

Since 2012, monthly cases of Middle East respiratory syndrome coronavirus (MERS-CoV) continue to cause severe respiratory disease that is fatal in ~35% of diagnosed individuals. The ongoing threat to global public health and the need for novel therapeutic countermeasures have driven the development of animal models that can reproducibly replicate the pathology associated with MERS-CoV in human infections. The inability of MERS-CoV to replicate in the respiratory tracts of mice, hamsters, and ferrets stymied initial attempts to generate small animal models. Identification of human dipeptidyl peptidase IV (hDPP4) as the receptor for MERS-CoV infection *opened the door* for genetic engineering of mice. Precise molecular engineering of mouse DPP4 (mDPP4) with clustered regularly interspaced short palindromic repeats (CRISPR)/Cas9 technology maintained inherent expression profiles, and limited MERS-CoV susceptibility to tissues that naturally express mDPP4, notably the lower respiratory tract wherein MERS-CoV elicits severe pulmonary pathology. Here, we describe the generation of the 288–330$^{+/+}$ MERS-CoV mouse model in which mice were made susceptible to MERS-CoV by modifying two amino acids on mDPP4 (A288 and T330), and the use of adaptive evolution to generate novel MERS-CoV isolates that cause fatal respiratory disease. The 288–330$^{+/+}$ mice are currently being used to evaluate novel drug, antibody, and vaccine therapeutic countermeasures for MERS-CoV. The chapter starts with a historical perspective on the emergence of MERS-CoV and animal models evaluated for MERS-CoV pathogenesis, and then outlines the development of the 288–330$^{+/+}$ mouse model, assays for assessing a MERS-CoV pulmonary infection in a mouse model, and describes some of the challenges associated with using genetically engineered mice.

Key words Middle East respiratory syndrome coronavirus, Mouse, Clustered regularly interspaced short palindromic repeats, Cas9, Pathogenesis

1 Introduction

In February of 2018 MERS-CoV was listed as a priority on the R&D Blueprint for the global strategy and preparedness plan outlined by the World Health Organization (WHO) [1]. The R&D Blueprint includes viruses that pose a global public health risk, and for which there are no available therapeutic countermeasures [1]. Twenty-seven countries have reported cases of MERS-CoV

Rahul Vijay (ed.), *MERS Coronavirus: Methods and Protocols*, Methods in Molecular Biology, vol. 2099,
https://doi.org/10.1007/978-1-0716-0211-9_12,

with most cases confined to the Arabian Peninsula. Diagnosed cases of MERS-CoV in countries outside the Arabian Peninsula are primarily traveler associated. The potential for global spread of MERS-CoV was realized in 2015 when a single traveler returning to South Korea initiated an outbreak that infected 186 people resulting in 20% fatality and caused widespread fear that crippled the economy for nearly 6 months [2–4]. Human-to-human transmission is often associated with close contact in the health care setting, but can also occur between family members within a household [5]. Asymptomatic individuals pose a particular risk of transmission due to their unknown carrier status as demonstrated in the health care setting [6]. Despite the high percent of fatalities associated with MERS-CoV outbreaks on the Arabian Peninsula most epidemiological studies suggest R_0 values <1, indicative of a low risk of sustainable human-to-human transmission, whereas epidemiological studies from the South Korean outbreak describe R_0 values (>1) akin to more sustainable human-to-human transmission [7]. Recurring spillover events from dromedary camels (zoonotic reservoir for MERS-CoV on the Arabian Peninsula) likely contribute to newly diagnosed cases in humans [8–10]. The potential for continuous reintroduction to humans increases the risk of MERS-CoV adapting in humans to acquire enhanced human-to-human transmission profiles, a scenario suspected to have initiated the SARS-CoV pandemic in 2002–2003 [11]. Effective public health measures and culling of civet cats, the zoonotic host for SARS-CoV, brought the SARS-CoV pandemic to a rapid end [11]. Eliminating MERS-CoV through culling of infected camel herds is not a practical solution. Furthermore, detection of pre-emergent MERS-CoV-like, and SARS-CoV-like, strains circulating in bat species indicate that the natural environment is ripe for future human exposures to potentially pathogenic coronaviruses [12–14]. Therefore, the development of therapeutic countermeasures that can interfere with MERS-CoV pathogenesis is critical to break zoonotic-to-human and human-to-human transmission cycles that may instigate global spread.

Evaluating the toxicity and efficacy of novel MERS-CoV therapeutics require the availability of animal models that effectively recapitulate MERS-CoV pathogenesis during fatal cases of human infections. Therefore, the first question in generating a MERS-CoV animal model would be: What are the pathological features of a human infection? Limited histopathological findings from human autopsies indicate that fatal cases of MERS-CoV results from pneumonia initiated by infection of bronchiolar and alveolar epithelia of the lower respiratory tract (LRT) [15, 16]. Pneumonia in the LRT is also the prominent finding on radiographs from X-rays and CTs of diagnosed human cases [17]. High viral loads in tracheal aspirates from patients are also associated with severe pulmonary disease [18], which is indicative of actively replicating MERS-CoV in the LRT. Initial evaluation of the human MERS-CoV EMC/2012

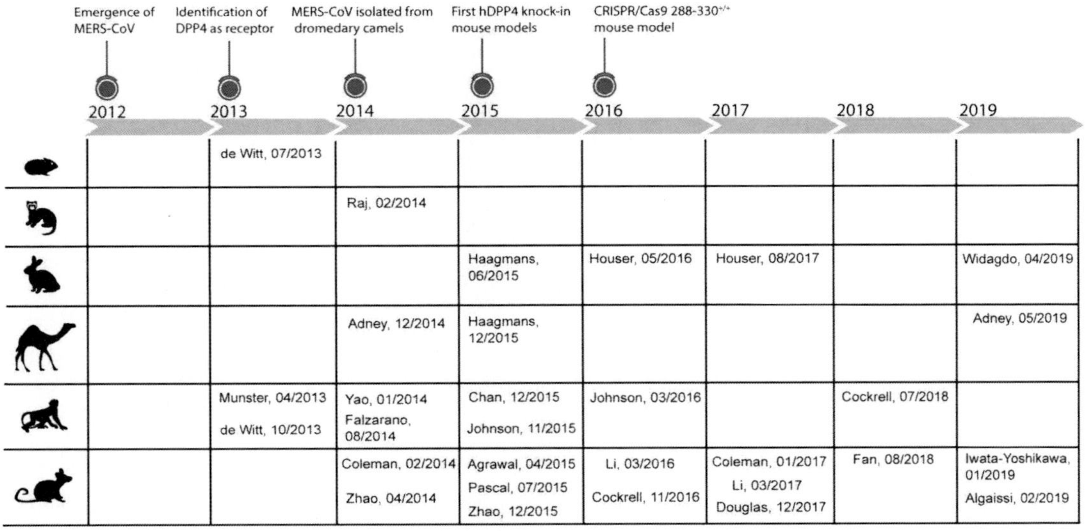

Fig. 1 Timeline of the mammalian models evaluated for MERS-CoV pathogenesis between 2012 and 2019. Specific events since the emergence of MERS-CoV in 2012 are emphasized above the timeline. References to mammalian models evaluated for MERS-CoV pathogenesis comprise hamster [19], ferret [20], rabbit [21–24], camel [25–27], nonhuman primates [28–35], and mouse [36–48]

isolate in rhesus macaques demonstrated replication in the LRT with mild pneumonia-like disease (Fig. 1) [28]. Achieving respiratory pathology reflecting a lethal human disease proved to be more complicated in nonhuman primates. Severe respiratory disease in the marmoset produced clinical endpoints consistent with fatal disease that required euthanasia (Fig. 1) [29, 30]. Evaluation of two human isolates, Jordan and EMC/2012, and a tissue culture-adapted MERS-CoV strain (MERS-0) in nonhuman primates resulted in mild disease in rhesus macaques or marmosets (Fig. 1) [31–33], confounding the reproducibility of near-lethal disease in NHPs. Nonhuman primates are central to late-stage preclinical evaluation of therapeutic countermeasures, but may be impractical for initial preclinical studies. A small animal model may be applicable if there is limited therapeutic available for toxicity and efficacy testing, especially if large animal numbers are needed to determine confidence and reproducibility.

Early studies in mouse, hamster, and ferret revealed that conventional small animal models were fully resistant to MERS-CoV infection and replication (Fig. 1) [19, 20, 36]. A seminal study identifying the MERS-CoV receptor as human dipeptidyl peptidase IV (hDPP4) [49], and publication of the crystal structure of hDPP4 interacting with the receptor binding domain (RBD) of the MERS-CoV spike protein [50], exposed tropism determinants critical for susceptibility. Dipeptidyl peptidase IV contact amino acids at the hDPP4/RBD interface are highly conserved among MERS-CoV-susceptible mammalian species (human, camel, and

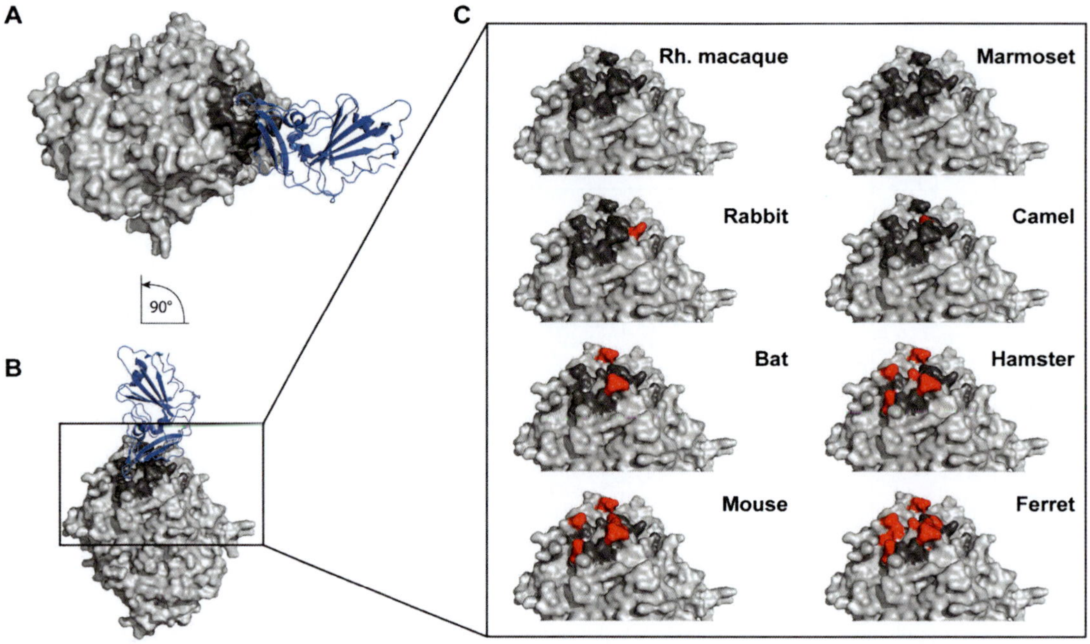

Fig. 2 Comparison of DPP4 from different species. (**a**) Horizontal view of the crystal structure (PDB, 4L72) of human DPP4 (light gray) interacting with the MERS-CoV receptor binding domain (RBD; blue). The contact residues of human DPP4 with MERS-CoV RBD are highlighted in dark gray. (**b**) A 90° rotation, demonstrating the vertical view of (A). (**c**) Zoomed-in view of the human DPP4 structure (light gray) with highlighted MERS-CoV RBD contact residues (dark gray). Species-specific contact residues that differ from human are highlighted in red

bat) (Fig. 2) [51]. Although mouse, hamster, ferret, and guinea pig DPP4 orthologs exhibit high overall similarity to hDPP4, specific amino acid differences at the DPP4/RBD interface account for the inability of these species to support infection [51–56]. Overexpression of a mouse DPP4 (mDPP4) with changes in the contact residues at the DPP4/RBD altered cellular profiles from resistant to susceptible to MERS-CoV infection [52, 53, 56]. The dependence on DPP4-specific contact was further substantiated by similar studies evaluating modified DPP4 orthologs from the hamster, ferret, and guinea pig [55]. Dipeptidyl peptidase IV was identified as the major determinant of MERS-CoV tropism.

1.1 MERS-CoV Mouse Models

Researchers rapidly leveraged knowledge of the DPP4 receptor to generate susceptible small animal models (Fig. 1) [12]. Zhao et al. utilized a unique approach for producing susceptible mice that could replicate human isolates of MERS-CoV in the lungs by infecting mouse lungs with an adenovirus that constitutively expresses the full-length hDPP4 gene (Fig. 1) [37]. Transient expression of hDPP4 supported infection and replication with human strains of MERS-CoV in the lungs and indicated that this technology may be an effective rapid response platform for initial evaluation of emergent and pre-emergent viruses. However, pathology associated with a fatal MERS-CoV infection was not observed in the Ad-hDPP4 model [37], which limited the capacity to evaluate the efficacy of therapeutic countermeasures.

Genetic engineering of mice would be necessary to develop preclinical MERS-CoV mouse models with respiratory phenotypes that reflected clinical outcomes in patients. Knock-in of full-length hDPP4 rendered mice susceptible to human isolates of MERS-CoV at low infection doses (Fig. 1) [38–40]. Knock-in mice exhibited severe pulmonary pathology and increased mortality; however, widespread constitutive expression of full-length hDPP4 resulted in high levels of MERS-CoV infection and replication in extrapulmonary tissues [38–40]. In some studies, higher viral loads could be detected in the brain compared to the lungs [39, 40]. Mice with infections of the central nervous system (CNS) exhibited encephalitis that corresponded with the kinetics of mortality [39]. Currently, there is no evidence to support a CNS component associated with MERS-CoV pathogenesis in humans. Attempts to restrict hDPP4 expression to epithelial cells of the lungs using constitutive tissue specific promoters (e.g., cytokeratin K18) yielded outcomes similar to those observed with SARS-CoV mouse models, wherein high levels of MERS-CoV infection/replication were detected in the brains (Fig. 1) [39].

To circumvent confounding problems associated with global bio-distribution of overexpressed hDPP4 receptor, researchers engineered mouse models using sophisticated molecular approaches. Pascal et al. employed Regeneron's VelociGene technology to replace sequences encoding nearly the entire mDPP4 genomic region with those encoding the exons/introns from the hDPP4 genetic region (Fig. 1) [41]. Retaining the mDPP4 5′ and 3′ genetic elements that regulate expression maintained inherent expression profiles of full-length hDPP4 in mice [41]. Importantly, MERS-CoV infection/replication was readily detected in the lungs with little involvement of extrapulmonary tissues [41]. Infection with human isolates of MERS-CoV caused moderate respiratory pathology with mortality determined by euthanasia of mice at 20% weight loss [41]. Unfortunately, commercial restrictions limit the availability and use of this model to the broader scientific community. In addition to the concerns raised above, the first generation of mouse models was developed with the full-length hDPP4, which may alter the inherent physiological properties of the mouse.

The multifaceted involvement of DPP4 in maintaining immune homeostasis is of significant importance regarding susceptibility to infectious disease [57]. DPP4 exists in two forms: (1) a membrane anchored form on the surface of multiple cells types (e.g., B cells, T cells, NK cells, and epithelial cells to mention a few) and (2) a secreted form that can be identified in human serum [57]. DPP4 interacts with and modifies heterologous protein molecules involved in nociception, neuroendocrine function, metabolism, cardiovascular function, immune regulation, and infection [57]. Modification of heterologous protein function can proceed through cleavage of *N*-terminal amino acids through the enzymatic activity of the α/*b*-hydroxylase domain, or allosteric interaction/signal transduction [57]. The species specificity of DPP4 is exemplified by the interaction of hDPP4 with adenosine deaminase (ADA), a well-recognized binding partner of hDPP4, which modulates downstream T cell functions [58–60]. The hDPP4/ADA interaction evolved in higher mammalian species (human, NHP, bovine, rabbit), but not in mouse or rat [58–60]. Interestingly, in one study ADA was demonstrated to block infection of MERS-CoV in tissue culture [20], indicating that the binding site on hDPP4 for ADA, and the MERS-CoV RBD, may overlap. Consequently, introducing full-length hDPP4 into mice may skew innate immune mechanisms that could influence responses to therapeutic countermeasures.

In the second generation of MERS-CoV-susceptible mouse models amino acid residues predicted to function at the mDPP4/MERS-CoV RBD interface were modified to avoid the introduction of full-length hDPP4 (Fig. 1) [42, 43]. Li et al. recently developed a mouse model wherein the *mDPP4* genomic region encompassing exons 10–12 were replaced with the respective genomic region from *hDPP4*, referred to as an hDPP4 knock-in model (hDPP4-KI) [43]. Exons 10–12 encode contact amino acids at the hDPP4/MERS-CoV RBD interface that were able to support replication of human MERS-CoV isolates in the lungs, but did not elicit a mortality phenotype [43]. Adaptive evolution of human MERS-CoV in the hDPP4-KI mouse resulted in mouse-adapted viruses that evoked a lethal respiratory phenotype with little involvement of extrapulmonary tissues. The lethal respiratory phenotype is a consequence of novel mutations acquired during adaptive evolution. A combination of mutations in both the S1 and S2 regions of the MERS-CoV spike protein facilitated a lethal respiratory phenotype [43]. Results in the hDPP4-KI model substantiate an earlier mouse model referred to as the $288–330^{+/+}$ model, which was designed with only two amino acid changes in mDPP4 to generate MERS-CoV susceptible mice.

Genetic engineering and implementation of the $288–330^{+/+}$ mouse model, combined with MERS-CoV adaptive evolution, is the subject of this chapter. Initial studies in tissue culture revealed that human and rodent cell types were resistant to MERS-CoV infection upon overexpression of mDPP4; however, overexpression of hDPP4 conferred permissivity to infection/replication [53]. Comparative structural modeling of hDPP4 and mDPP4 revealed putative contact residues in mDPP4 amenable to modification at the DPP4/RBD interface. Modification of two amino acids (A288L and T330R) was sufficient to endow mDPP4 with the capacity to mediate MERS-CoV infection/replication [53]. Shortly after the emergence of MERS-CoV into humans in 2012, the CRISPR/Cas9 genome editing technology became available for applications to modify mammalian genomes in vitro and in vivo [61–63]. Recognizing our unique situation, we designed CRISPR/Cas9 targets to modify the mouse genome encoding amino acids A288 and T330 in exons 10 and 11 of the *mDPP4* gene (Fig. 3) [12, 42]. Concomitant with mouse development, in vitro studies were initiated to adapt MERS-CoV to the modified mDPP4 [42]. Tissue culture adaption resulted in MERS-0 virus, which contained an RMR insertion and S885L mutation in the S2 region of the MERS-CoV spike protein [42]. A MERS-0 molecular clone exhibited enhanced replication kinetics and higher titers compared to human MERS-CoV isolates. Additionally, the MERS-0 virus replicated to higher levels in the lungs of $288–330^{+/+}$ mice, compared to human and camel MERS-CoV isolates [42]. Based on these data the MERS-0 virus was used to initiate passaging in mice heterozygous for mDPP4 with A288L and T330R mutations, $288–330^{+/-}$ (Fig. 4). We reasoned that adaptation around one expressed copy of the mDPP4 with 288–330 mutations, and a wild-type mDPP4 expressed copy, might cultivate generation of a mouse-adapted MERS-CoV that could utilize wild-type mDPP4 as the primary receptor. After 15 passages we obtained a mouse-adapted MERS-CoV (MERS15c2) exhibiting a lethal respiratory phenotype in the $288–330^{+/+}$ mice [42]. Our MERS-CoV reverse genetic system was used to generate an infectious clone of the mouse-adapted virus, icMERSma1 [42]. Lethal respiratory pathology with icMERSma1 required high infectious doses (5×10^6 Pfu). An additional 20 passages of icMERSma1 in $288–330^{+/-}$ mice bore a novel mouse-adapted MERS-CoV that produced lethal respiratory disease at doses of 5×10^5 Pfu, and lung pathology associated with severe respiratory disease at 5×10^4 to 5×10^5 Pfu [44] (Fig. 1). This MERS-CoV model system ($288–330^{+/+}$ mice and mouse-adapted MERS-CoV viruses) is now being employed to: (1) understand complex virus-host interactions [12, 31, 42, 64–67], (2) evaluate antibody-based therapeutics [42], (3) evaluate drug-based therapeutic countermeasures [68], and (4) evaluate anti-MERS-CoV vaccines [42, 66]. The goal of this chapter is to provide an outline of how to rationally design a mouse with altered susceptibility to MERS-CoV. For additional information there are a number of detailed reviews and book chapters describing the design and utilization of the CRISPR/Cas9 technology for generating mouse models [69, 70].

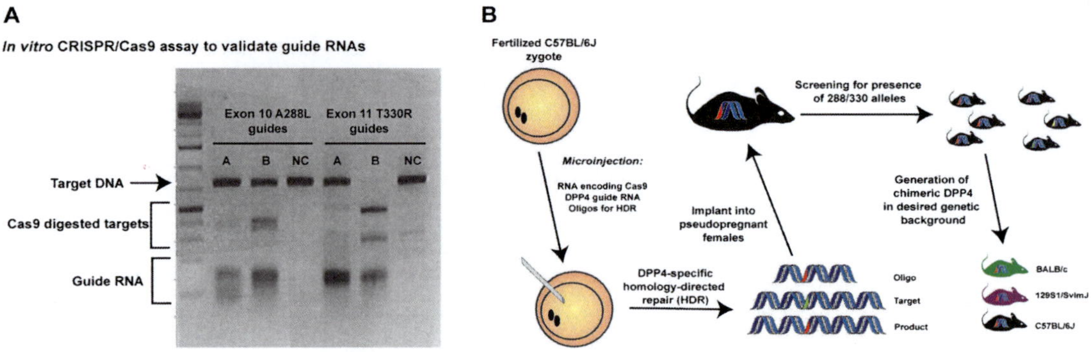

C

Cas9 guide sequence	Oligo for Homology Directed Repair (HDR)	Engineered amino acid
5'-gCTCCTgCATCTgTggCAAg-3'	5'-TCAgCTCATCCTCTAgTgCggCTCCCATCCAAATCC CTgCTCCTgCATCTgTgTTgAgAgggTAAgATgCTCTAC CTgAgTgCTTgggTgTAgATCgCTTTCCACTAC-3'	Ala288Leu
5'-ATgATAAgATCAACCTAACg-3'	5'-gATTCAgAACTATTCCgTgATggCTATCTgTgACTAT gATAAgATCAACCTAAgATggAACTgTCCATCCATCC gTAAgATTTCCCTgggATgCTCTgCTgTTCCAACAgCA-3'	Thr330Arg

D

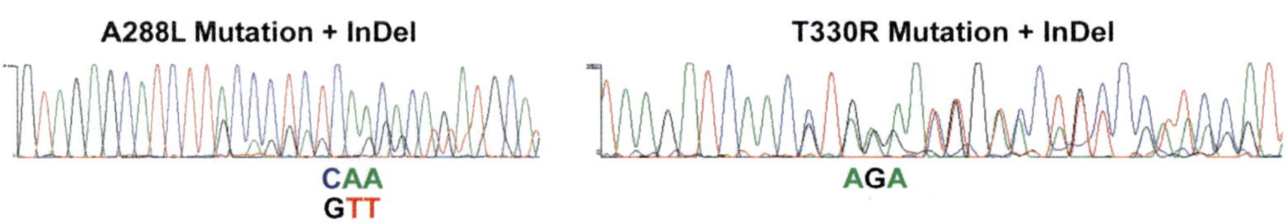

Fig. 3 CRISPR/Cas9 mediated genetic engineering of mouse DPP4. (**a**) In vitro validation of guide RNAs via Cas9 endonuclease assay (image was kindly provided by Dale Cowley in the Animal Models Core Facility at the University of North Carolina). Agarose gel separation based on size allows for discrimination between target DNA, Cas9 digested targets, and guide RNAs. (**b**) Schematic utilizing CRISPR/Cas9 technology to genetically engineer mice. Fertilized C57BL/6 J zygotes are collected and injected with RNA encoding Cas9, DPP4 single guide RNA, and oligos to facilitate homology-directed repair (HDR). Microinjected zygotes are implanted into pseudopregnant recipient female C57BL/6 J mice. Offspring are screened by sequencing for the intended change at positions 288 and 330. Mice identified as having the appropriate changes are backcrossed to C57BL/6 J mice to maintain the pure C57BL/6 J background, or may be crossed to any desired strain (e.g., BALB/cJ or 129S1/SvImJ). (**c**) Table describing sequences of Cas9 guide RNAs and oligos for HDR to genetically engineer amino acid changes at position 288 (Ala to Leu) and 330 (Thr to Arg). (**d**) Sequencing chromatograms highlighting how the F0 offspring from embryo implantation can be a mosaic of insertion/deletions (InDel's) generated by random non-homologous end joining from Cas9 cutting at the genomic alleles, and the HDR repair that incorporates the intended changes encoding amino acids at positions 288 and 330. Pure homozygous 288–330$^{+/+}$ lines were obtained by backcrossing onto C57BL/6 J mice. The highlighted mutations CAA (TTG in the reverse orientation) and AGA encode the novel 288 L and 330R amino acids

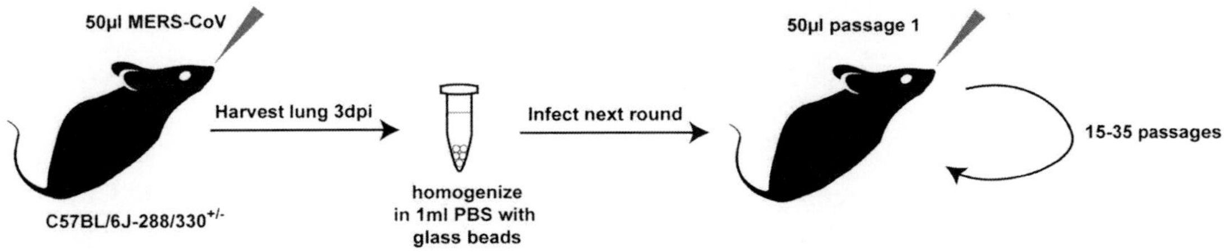

Fig. 4 Mouse adaptation of MERS-0 in 288–330$^{+/-}$ mice. 288/330$^{+/-}$ mice were intranasally infected with 50 μL of MERS-0. Three days after infection lungs were harvested, homogenized in 1 mL PBS with glass beads, and 50 μL of the supernatant from the lung homogenate was used to infect the next round of 288–330$^{+/-}$ mice. Serial lung passages are performed for 15 [42] to 35 [44] rounds

2 Materials

2.1 Tissue Culture Adaptation of MERS-CoV to mDPP4 Harboring the A288L and T330R Mutations

1. Mouse fibroblast NIH/3T3 cells (ATCC, cat# CRL-1658) stably expressing mDPP4 with A288L and T330R mutations [42].
2. NIH/3T3 maintenance in standard DMEM media supplemented with 10% FBS and 1× antibiotic/antimycotic.
3. MERS-CoV-tRFP recombinant virus generated using coronavirus reverse genetics [71].
4. See plaque assay materials for clonal isolation and expansion (Subheading 2.8).

2.2 Intranasal Infection of 288–330$^{+/+}$ Mice

1. Virus diluted in viral growth medium (OptiMEM, 3% Fetal Clone II, 1× antibiotic/antimycotic, 1× NEAA). Select a virus concentration that is appropriate for the specific study design.
2. Sterile 1 mL syringes and needles.
3. Ketamine/xylazine mixture (100 mg/mL stock solution, Akorn, Lake Forest, IL).
4. Scale capable of weighing mice ranging from 10 to 50 g.
5. Ear punch tool.

2.3 Monitoring Morbidity and Mortality

1. Scale capable of weighing mice ranging from 10 to 50 g.
2. Plastic cup to hold mice while weighing.
3. Plastic square to cover plastic cup.

2.4 Assessment of Respiratory Function as Additional Metric for Morbidity

1. Buxco system.

2.5 Harvest of Lungs from Infected Mice

1. Scale capable of weighing mice ranging from 10 to 50 g.
2. Sealable jar filled with paper towels and isoflurane (99.9% isofluorane, Piramal, Andhra Pradesh, India).
3. Styrofoam board.
4. Two 50 mL conical tubes filled with 70% ethanol and Cidecon.
5. Scissors, forceps, and metal pins.
6. Collection tubes prefilled with appropriate solutions.

2.6 Harvest of Bronchoalveolar Lavage (BAL) from Infected Mice

1. Styrofoam board.
2. Sealable jar filled with paper towels and isoflurane.
3. Two 50 mL conical tubes filled with 70% ethanol and Cidecon.
4. Surgical tools (scissors, forceps, and metal pins).
5. Collection tubes.
6. Safelet catheters 22G X1″(Exel International).
7. PBS filled 1 mL luer lock syringe.

2.7 Harvest of Blood and BAL for Hematological Analysis via VetScan HM5

1. Styrofoam board.
2. Sealable jar filled with paper towels and isoflurane.
3. Two 50 mL conical tubes filled with 70% ethanol and Cidecon.
4. Surgical tools (scissors, forceps, and metal pins).
5. EDTA prefilled collection tubes.
6. Safelet catheters.
7. PBS filled 1 mL luer lock syringe.

2.8 Plaque Assay of Lungs from Infected Mice

1. Homogenizer.
2. Centrifuge.
3. Vero CCL81 cells (ATCC, Manassas, VA) grown in appropriate medium (DMEM, 10% Fetal Clone II, 1× antibiotic/antimycotic).
4. 6-Well plates.
5. Two bottles (250 mL or 500 mL) for overlay.
6. Dilution tube boxes (96 tube format) filled with 450 μL PBS.
7. Overlay: one bottle with water and agarose (0.8 g/100 mL); one bottle with 2× medium (high glucose MEM, 20% Fetal Clone II, 2× antibiotic/antimycotic).
8. Microwave.
9. Bead bath (37 °C and 56 °C).
10. Incubator at 37 °C.

3 Methods

Important: All experiments using MERS-CoV strains should be executed under BSL3 conditions in accordance with Institutional Biosafety Committee approval, and under the governance of the National Institute of Allergy and Infectious Diseases within the National Institute of Health.

3.1 Tissue Culture Adaptation of MERS-CoV to mDPP4 Harboring the A288L and T330R Mutations (See Note 1)

1. Passaging was initiated with a MERS-CoV infectious clone generated using reverse genetics techniques to staple together sequences similar to the human EMC/2012 strain [71, 72] (*see* **Note 2**).
2. Mouse NIH 3T3 cells were generated to stably express the mDPP4 containing the A288L and T330R mutations [42] (*see* **Note 3**).
3. Seed NIH 3T3 cells stably expressing the modified mDPP4 receptor on 6-well plates at 1×10^6 cells/well.
4. 24 h after seeding, infect cells with the rMERS-CoV at an MOI of 0.1–0.001 (*see* **Note 4**).
5. If the virus is labeled with a fluorescent marker, cells are monitored for increased fluorescence at 24, 48, and 72 h post-infection using a fluorescent microscope (*see* **Note 5**).
6. Plaque-like islands of tRFP-expressing cells are indicative of replicating virus.
7. Harvest supernatant from infected cells at 48 to72 h. This is considered the passage 1 (P1) virus.
8. The passaging cycle is continued by diluting the supernatant 1:100–1:1000 and repeating the infection on fresh NIH/3T3 cells stably expressing mDPP4 with the A288L and T330R mutations.
9. After a predetermined number of passages the region encoding the spike protein of MERS-CoV is sequenced using RT-PCR to amplify the region of interest followed by standard Sanger sequencing (*see* **Note 6**).
10. After 10 passages viruses were plaque purified by diluting the heterogenous stock of virus $10^{-1}$to 10^{-6}, and infecting a monolayer of Vero CCL81 cells similar to a standard plaque assay.
11. Single plaques are isolated using a pipet tip and the virus expanded on a freshly seeded monolayer of Vero CCL81 cells.
12. Virus stocks are generated, viral RNA is isolated using standard TRIzol purification, and the region encoding the MERS-CoV spike protein is amplified by standard RT-PCR techniques and sequenced using standard Sanger sequencing.

13. Mutations identified by sequencing must be confirmed using a reverse genetic system to generate an infectious clone encoding the identified mutations [72]. Cockrell et al. validated the MERS-0 virus in this manner [42].

3.2 Engineering the 288–330$^{+/+}$ Mouse Model with CRISPR/Cas9 Homology-Directed Repair Genome Editing (Fig. 3)

The details for generating and using the CRISPR/Cas9 system to generate mutations are outlined in the materials and methods by Cockrell et al. [42]. Notably, the 288–330$^{+/+}$ mice were initially generated in the Animal Models Core facility at the University of North Carolina at Chapel Hill. The extensive technical expertise required for genetic engineering of mice is the subject of many expert reviews and book chapters that will not be covered here. Nevertheless, we provide a conceptual overview of the steps to generate the 288–330$^{+/+}$ mice.

1. Design guide RNAs to target each of the A288 and T330 alleles. Cockrell et al. designed the guide RNAs to direct the Cas9 to cut as near the mutation site as possible (Fig. 3) (*see* **Note 7** for helpful resources to design and genetically engineer mouse knockouts).
2. Test guide RNAs in vitro for the capacity to cut a target sequence (Fig. 3).
   (a) Generate double-stranded ODNs or a plasmid containing the target sequence with the correct PAM site.
   (b) Assemble ribonucleoprotein (RNP) complexes according to manufacturer's instructions (*see* **Note 8**).
   (c) Subject double-stranded DNA with target sequence to RNP complexes and assess digestion pattern on an agarose gel (Fig. 3).
3. Two separate oligos are also designed to introduce the novel mutations on exons 10 (A288L) and 11 (T330R) of mDPP4, through homology-directed repair (Fig. 3).
4. Fertilized zygotes are collected from C57BL/6J female mice that were superovulated and mated to male C57BL/6J mice.
5. In vitro transcribed RNAs encoding the guide sequences and Cas9 endonuclease, combined with ODNs encoding the replacement alleles for 288 and 330 in mDPP4, were all pronuclear injected into the fertilized zygote [42] (Fig. 3). The fertilized zygotes were from C57BL/6J mice.
6. The injected embryos were implanted into pseudopregnant recipient females.
7. Newly born pups were screened for the presence of the correct change at the 288 and 330 alleles by standard PCR amplification and Sanger sequencing (Fig. 3) (*see* **Note 9**).

8. Mice identified to have both mutations were crossed with wild-type C57BL/6J mice to obtain heterozygous mice with the mutated 288 L and 330R alleles in *cis*.
9. The F1 heterozygous mice were intercrossed to generate homozygous breeders to develop the $288–330^{+/+}$ colony, and for utilization in subsequent experiments.
10. The mouse colony is maintained under conditions delineated by the institutional Department of Comparative Medicine (DCM) and Institutional Animal Use and Care Committee (IACUC).

Important: All experiments infecting mice with various MERS-CoV wild-type or recombinant isolates are performed under BSL3/ABSL3 conditions. The generation of recombinant isolates requires prior approval of the Institutional Biosafety Committee (IBC). Additionally, all animal experiments should have prior approval according to the Institutional Animal Use and Care Committee (IACUC), Institutional Biosafety Committee (IBC), and in accordance with the recommendations for the care and use of animals by the Office of Laboratory Animal Welfare at NIH.

3.3 Intranasal Infection of $288–330^{+/-}$ and $288–330^{+/+}$ Mice

1. Mice are randomly assigned into cohorts specifically paying attention to include age- and sex-matched control animals.
2. Individual mice are made identifiable (e.g., ear notch, ear clip, tail mark).
3. $288–330^{+/+}$ mice are brought into the BSL3 laboratory 7 days before start of the experiment to allow for environmental acclimation.
4. Mice are anesthetized via intraperitoneal injection of 50–100 μL of a ketamine (50 mg/kg)/xylazine (15 mg/kg) mixture (*see* **Note 10**).
5. Level of anesthesia is assessed by pedal reflex.
6. Measure initial weight (Day 0 weight) while waiting for mouse to be anesthetized (*see* **Note 11**).
7. Once a pedal reflex is no longer triggered, mice should be immediately infected by the intranasal route. Holding the animal vertically, apply 50 μL of virus solution by pipetting onto their nostrils and allow them to inhale. To ensure that all of the 50 μL reach the lower respiratory tract hold the mouse upright for an additional 30 s (*see* **Note 12**).
8. Note any inconsistencies during infection, including: (1) presence of bubbles of inoculum from nasal cavity, (2) occurrence of inoculum in mouth, or (3) failure to inhale entire dose of inoculum. Notes will help to explain potential inconsistencies in readout parameters and may be used as exclusion criteria for inefficient infections.

9. Mice are put back into the cage and placed on their back to ensure virus solution will stay in the lungs. Note: Cages are returned to cage rack, but the respiration of mice is continuously monitored by observing breathing.
10. Place mice next to each other to keep body temperature as close to normal as possible.
11. Check cage after 30–45 min to confirm that all mice wake up from anesthesia and infection.

3.4 Mouse Adaptation of MERS-0 in 288–330$^{+/-}$ Mice (Fig. 4)

1. Mouse adaptation was initiated in heterozygous 288–330$^{+/-}$ mice by infecting with 50 μL of the MERS-0 infectious clone (*see* **Note 13**).
2. At 3 days post-infection the mouse is euthanized by extended exposure to isoflurane. ~2 mL of isoflurane is added to the bottom of a jar that can be firmly sealed (*see* **Note 14**).
3. A thoracotomy is performed to expose the lungs.
4. Lungs are removed and placed in a 2 mL gasket sealed skirted screw cap tubes. Tubes are previously prepared with 1 mL of 1× PBS containing ~5–10 mm of glass beads.
5. Lungs are homogenized for 60 s in a bead homogenizer.
6. Lung lysates are centrifuged in a microcentrifuge for 5 min at max speed to pellet debris.
7. This is considered passage 1 (P1) and 50 μL of lung homogenate is used to infect a naïve 288–330$^{+/-}$ mouse.
8. The process is repeated for a desired number of cycles.

3.5 Monitoring Morbidity and Mortality

1. After infection mice are monitored daily for weight loss for the entire duration of the experiment.
2. To record daily weights, pick up mice by the tail, identify by ear notch, and place into cup on a scale. Record weight and calculate percentage of starting body weight (*see* **Note 15**).
3. Mice can also be monitored to determine if they are moribund using a clinical scoring scale whereby: 0 = no clinical signs; 1 = ruffled fur; 2 = ruffled fur with hunched posture (only slight with no signs of dehydration); 3 = as defined in number 2 with more severe signs of dehydration and some loss of body strength; 4 = pronounced dehydration and prominent loss of mobility; 5 = unresponsive to stimuli and prominent eye squinting.
4. It is important to note that weight loss might not always be the most appropriate parameter and animals should be euthanized at the discretion of the researcher even if animals have not reached 80% of their starting weight.

5. Mice that approach 80% of their starting body weight (20% weight loss) are euthanized via isoflurane overdose followed by a secondary euthanasia method (thoracotomy or cervical dislocation). Depending on the experimental circumstances an institutionally approved exception may be implemented to allow continuation of the experiment (increasing the frequency with which the mice are monitored will likely need to be implemented) (*see* **Note 16**).

3.6 Assessment of Respiratory Function as an Additional Metric for Morbidity (A Brief Stepwise Overview)

1. Acclimate individual mice for 30 min in plethysmography chamber (Buxco System, Wilmington, NC). A study by Menachery et al. comprehensively describes the application of plethysmography to infectious respiratory viruses in mouse models [73] (*see* **Note 17**).
2. A variety of lung function-related parameters are then recorded over a period of 5 min [e.g., enhanced pause (PenH), mid-tidal expiratory flow (EF50), peak expiratory flow (PEF), and peak inspiratory flow (PIF)].
3. Respiratory function can be performed every day over the entire course of infection, or on single selected days. Investigating a novel respiratory virus may require the investigator to perform a time course to determine the most effective time points for measurement. The largest differences between groups typically correlate with peak viral replication.
4. At each time point measured the mice need to be randomized into different chambers to avoid technical artifacts (e.g., a mouse measured at day 1 in chamber 1 should be evaluated in a different chamber on day 2 measurement). Practical considerations dictate that 8–12 animals can be measured at any one time, and each group of 8–12 mice may take an hour for a proficient technician. Therefore, experiments should be carefully planned to limit the number of mice to be evaluated.
5. Mice that are difficult to handle can be slightly anesthetized by applying isoflurane to the chamber in order to remove them from the chamber and return to their cage (*see* **Note 18**).

3.7 Measuring Lung Hemorrhage for Gross Assessment of Pathology and Harvesting the Lungs

Important: Removal of all samples from BSL3 facilities must be executed in accordance with the Institutional Biosafety Committee. Assays must either be performed inside the BSL3 laboratory or additional processing steps (validated to fully inactivate virus and approved by the IBC) must be executed before removal of samples to a BSL2 setting.

1. Sacrifice mice by isoflurane overdose.
2. Place on scale and record endpoint weights.

3. Pin mouse on all four extremities to Styrofoam board using metal pins.
4. Spray with 70% ethanol to avoid contamination of samples with fur.
5. Remove fur over thorax.
6. Open thorax, paying attention not to cut into lung tissue.
7. Assess lung tissue for reddish discoloration and record severity by applying a number 0 (no hemorrhaging) to 4 (severe hemorrhaging in all lobes of the lungs).
8. Harvest lung tissue and place in tubes prefilled with sample type specific solution.
9. Put scissors and forceps first into Cidecon to remove any residual blood and then into 70% ethanol in order to not cross-contaminate samples.
10. Whole lung can be used for one assay or different lobes can be used for different assays.

3.8 Collecting Blood from Infected Mice

1. **Steps 1–6** from Subheading 3.7.
2. Cut into superior vena cava and collect blood with a pipette.
3. Blood is typically transferred to a serum/plasma separation tube that allows for separation of serum/plasma from cells.
4. In the event that it is necessary to harvest cells for flow cytometry analysis or VetScan HM5 analysis, the blood sample can be transferred to a tube containing EDTA. Note: VetScan HM5 is a veterinary diagnostic machine that analyzes basic immune cell counts and additional hematological parameters within 2 min.
5. All VetScan assays are performed under BSL3 conditions. Removal of samples for downstream analysis outside of BSL3 conditions require specific inactivation procedures that must be pre-approved by the institutional biosafety committee.

3.9 Collecting Bronchoalveolar Lavage (BAL) from Infected Mice

1. **Steps 1–6** from Subheading 3.7.
2. Expose trachea and enter with catheter.
3. Remove needle part of catheter and attach 1 ml luer lock syringe prefilled with 1 mL of 1× PBS.
4. Carefully inject 1 mL PBS into lungs, wait for 30 s, and pull 1× PBS back out. This is the BAL sample.
5. Place sample into a fresh collection tube.
6. Use a new catheter for every mouse.
7. In the event that it is necessary to harvest cells for flow cytometry analysis or VetScan HM5 analysis, the BAL sample can

be transferred to an empty tube or a tube containing EDTA, respectively.

8. All VetScan assays are performed under BSL3 conditions. Removal of samples for downstream analysis outside of BSL3 conditions require specific inactivation procedures that must be pre-approved by the institutional biosafety committee.

3.10 Plaque Assay of Lungs from Infected Mice (Performed Under BSL3 Conditions)

1. One day prior to performing the assay, seed 5 × 10^5 Vero CCL81 cells into each well of a 6-well plate (one plate per sample).
2. Prepare overlay: 2× medium + agar in water (1:1). You will need 2 mL overlay per well.
3. Place 2× medium bottle into 37 °C bead bath.
4. Heat 0.8% agar in water in a microwave and place at 56 °C.
5. Thaw lung samples in tubes filled with 1 mL of 1× PBS and glass beads.
6. Once thawed, place into tissue homogenizer for 60 s.
7. Centrifuge at 13,000 rpm (~15,000 × g) for 5 min.
8. Transfer 50 μL of sample into 450 μL of PBS. Mix samples well.
9. Perform serial tenfold dilutions (10^{-1} to 10^{-6}).
10. Transfer 200 μL of each dilution to individual wells on a 6-well plate.
11. Incubate for 1 h, rocking 6-well plates every 15 min.
12. Mix 2× medium with dissolved agarose.
13. Put 2 mL of overlay onto every well.
14. Place in 37 °C incubator for 3 days.
15. Virus plaques are visualized by neutral red staining (2 mL/well) and using a lightbox. Count plaques to determine the number of plaque forming units per milliliter (Pfu/mL).

4 Notes

1. It is not necessary to initiate MERS-CoV adaptation in tissue culture first, as was demonstrated by Li et al. [43]. Nonetheless, Cockrell et al. chose to initiate their MERS-CoV adaptation studies on tissue culture cells while the 288–330$^{+/+}$ mice were in the process of being generated [42]. Therefore, for the purposes of this chapter this is included as a potential starting point for MERS-CoV adaptation.
2. Cockrell et al. initiated adaptation experiments with a rMERS-CoV expressing the tomato red protein in place of the Orf5 ORF [42, 71, 72].

3. This can also be achieved using a readily transfectable human embryonic fibroblast cell line such as HEK293T cells and selecting for stably transfected cells.
4. Cells can be counted by seeding an extra well, but it is safe to assume that the cell number is approximately doubled after 24 h.
5. This can also be achieved using a microscope to assess plaque size if plaques can be readily detected.
6. Since the major determinant of MERS-CoV tropism is the spike protein, it would be anticipated that mutations having the most significant impact on infection might occur within the gene encoding the spike protein. Nonetheless, as described by Cockrell et al., Douglas et al., and Li et al., a number of mouse-adapted mutations were identified in genetic regions outside of the spike gene, which may have a significant influence on virus fidelity and evasion of host immune responses [42–44].
7. At the time that these mice were being generated in early 2014, CRISPR/Cas9 reagents were not readily available. Additionally, there were few bioinformatics tools available to facilitate guide RNA design and off-target potential. In the current research environment CRISPR/Cas9 reagents can be sourced from multiple commercial entities and there are a number of bioinformatics tools to assist with design. Addgene is a non-profit plasmid repository where CRISPR reagents and resources can be readily obtained (https://www.addgene.org/). Additional guidance for generating mice using CRISPR/Cas9 technology can be found in more comprehensive protocols [69, 70].
8. All relevant reagents and protocols can now be obtained from commercial sources as readily synthesized RNAs and purified proteins (e.g., Integrated DNA Technologies).
9. Although a number of pups may be identified to have the correct mutation, many will likely be mosaic for random mutations including insertions/deletions due to the higher efficiency of non-homologous end joining (NHEJ) after Cas9 digestion compared to the desired HDR employed to mediate allele modification.
10. The administered dose will depend on the weight of the animal which should be predetermined the day prior to initiating the experiment.
11. It is not necessary to anesthetize mice for measuring daily weights.
12. It is important that the inoculum reaches the lower respiratory tract for a successful MERS-CoV infection.
13. Mouse adaptation can be initiated with any MERS-CoV strain that exhibits some pulmonary replication.

14. Mice should never come into contact with the isoflurane. To prevent direct contact, a layer of aluminum foil is placed at the bottom of the jar and this is covered with two additional layers of paper towel.
15. The percent body weight is typically calculated after leaving the BSL3 environment. Therefore, the weight sheets should have the anticipated weights of each animal at 20% weight loss. This will provide a real-time indication of when the mice are approaching the criteria established for humane euthanasia.
16. Institutional approval is required for animals to be placed under exception. It cannot be emphasized enough that all animal work should be pre-approved by appropriate University IACUC and IBC committees and should be in accordance with the recommendations for the care and use of animals by the Office of Laboratory Animal Welfare at NIH.
17. Assessing respiratory function using plethysmography under BSL3 conditions requires costly equipment and extensive training prior to use.
18. Anesthesia should be avoided prior to measuring lung function to prevent interference with lung function measurements.

References

1. Mehand MS, Al-Shorbaji F, Millett P, Murgue B (2018) The WHO R&D Blueprint: 2018 review of emerging infectious diseases requiring urgent research and development efforts. Antivir Res 159:63–67
2. Bernard-Stoecklin S, Nikolay B, Assiri A, Bin Saeed AA, Ben Embarek PK, El Bushra H, Ki M, Malik MR, Fontanet A, Cauchemez S, Van Kerkhove MD (2019) Comparative analysis of eleven healthcare-associated outbreaks of Middle East respiratory syndrome coronavirus (Mers-Cov) from 2015 to 2017. Sci Rep 9:7385
3. Joo H, Maskery BA, Berro AD, Rotz LD, Lee YK, Brown CM (2019) Economic impact of the 2015 MERS outbreak on the Republic of Korea's tourism-related industries. Health Secur 17:100–108
4. Lee SI (2015) Costly lessons from the 2015 Middle East respiratory syndrome coronavirus outbreak in Korea. J Prev Med Public Health 48:274–276
5. Hui DS, Azhar EI, Kim Y-J, Memish ZA, M-d O, Zumla A (2018) Middle East respiratory syndrome coronavirus: risk factors and determinants of primary, household, and nosocomial transmission. Lancet Infect Dis 18: e217–e227
6. Alfaraj SH, Al-Tawfiq JA, Altuwaijri TA, Alanazi M, Alzahrani N, Memish ZA (2018) Middle East respiratory syndrome coronavirus transmission among health care workers: implication for infection control. Am J Infect Control 46:165–168
7. Park JE, Jung S, Kim A, Park JE (2018) MERS transmission and risk factors: a systematic review. BMC Public Health 18:574
8. Alshukairi AN, Zheng J, Zhao J, Nehdi A, Baharoon SA, Layqah L, Bokhari A, Al Johani SM, Samman N, Boudjelal M, Ten Eyck P, Al-Mozaini MA, Zhao J, Perlman S, Alagaili AN (2018) High prevalence of MERS-CoV infection in camel workers in Saudi Arabia. MBio 9
9. Conzade R, Grant R, Malik MR, Elkholy A, Elhakim M, Samhouri D, Ben Embarek PK, Van Kerkhove MD (2018) Reported direct and indirect contact with dromedary camels among laboratory-confirmed MERS-CoV cases. Viruses 10:E425
10. Hemida MG, Elmoslemany A, Al-Hizab F, Alnaeem A, Almathen F, Faye B, Chu DK, Perera RA, Peiris M (2017) Dromedary camels and the transmission of Middle East respiratory syndrome coronavirus (MERS-CoV). Transbound Emerg Dis 64:344–353

11. Peiris JS, Guan Y, Yuen KY (2004) Severe acute respiratory syndrome. Nat Med 10:S88–S97
12. Cockrell AS, Leist SR, Douglas MG, Baric RS (2018) Modeling pathogenesis of emergent and pre-emergent human coronaviruses in mice. Mamm Genome 29:367–383
13. Cui J, Li F, Shi ZL (2019) Origin and evolution of pathogenic coronaviruses. Nat Rev Microbiol 17:181–192
14. Menachery VD, Graham RL, Baric RS (2017) Jumping species-a mechanism for coronavirus persistence and survival. Curr Opin Virol 23:1–7
15. Alsaad KO, Hajeer AH, Al Balwi M, Al Moaiqel M, Al Oudah N, Al Ajlan A, AlJohani S, Alsolamy S, Gmati GE, Balkhy H, Al-Jahdali HH, Baharoon SA, Arabi YM (2018) Histopathology of Middle East respiratory syndrome coronovirus (MERS-CoV) infection—clinicopathological and ultrastructural study. Histopathology 72:516–524. https://doi.org/10.1111/his.13379
16. Ng DL, Al Hosani F, Keating MK, Gerber SI, Jones TL, Metcalfe MG, Tong S, Tao Y, Alami NN, Haynes LM, Mutei MA, Abdel-Wareth L, Uyeki TM, Swerdlow DL, Barakat M, Zaki SR (2016) Clinicopathologic, immunohistochemical, and ultrastructural findings of a fatal case of Middle East respiratory syndrome coronavirus infection in the United Arab Emirates, April 2014. Am J Pathol 186:652–658
17. Arabi YM, Balkhy HH, Hayden FG, Bouchama A, Luke T, Baillie JK, Al-Omari A, Hajeer AH, Senga M, Denison MR, Nguyen-Van-Tam JS, Shindo N, Bermingham A, Chappell JD, Van Kerkhove MD, Fowler RA (2017) Middle East respiratory syndrome. N Engl J Med 376:584–594
18. Oh MD, Park WB, Choe PG, Choi SJ, Kim JI, Chae J, Park SS, Kim EC, Oh HS, Kim EJ, Nam EY, Na SH, Kim DK, Lee SM, Song KH, Bang JH, Kim ES, Kim HB, Park SW, Kim NJ (2016) Viral load kinetics of MERS coronavirus infection. N Engl J Med 375:1303–1305
19. de Wit E, Prescott J, Baseler L, Bushmaker T, Thomas T, Lackemeyer MG, Martellaro C, Milne-Price S, Haddock E, Haagmans BL, Feldmann H, Munster VJ (2013) The Middle East respiratory syndrome coronavirus (MERS-CoV) does not replicate in Syrian hamsters. PLoS One 8:e69127
20. Raj VS, Smits SL, Provacia LB, van den Brand JM, Wiersma L, Ouwendijk WJ, Bestebroer TM, Spronken MI, van Amerongen G, Rottier PJ, Fouchier RA, Bosch BJ, Osterhaus AD, Haagmans BL (2014) Adenosine deaminase acts as a natural antagonist for dipeptidyl peptidase 4-mediated entry of the Middle East respiratory syndrome coronavirus. J Virol 88:1834–1838
21. Haagmans BL, van den Brand JM, Provacia LB, Raj VS, Stittelaar KJ, Getu S, de Waal L, Bestebroer TM, van Amerongen G, Verjans GM, Fouchier RA, Smits SL, Kuiken T, Osterhaus AD (2015) Asymptomatic Middle East respiratory syndrome coronavirus infection in rabbits. J Virol 89:6131–6135
22. Houser KV, Broadbent AJ, Gretebeck L, Vogel L, Lamirande EW, Sutton T, Bock KW, Minai M, Orandle M, Moore IN, Subbarao K (2017) Enhanced inflammation in New Zealand white rabbits when MERS-CoV reinfection occurs in the absence of neutralizing antibody. PLoS Pathog 13:e1006565
23. Houser KV, Gretebeck L, Ying T, Wang Y, Vogel L, Lamirande EW, Bock KW, Moore IN, Dimitrov DS, Subbarao K (2016) Prophylaxis with a Middle East respiratory syndrome coronavirus (MERS-CoV)-specific human monoclonal antibody protects rabbits from MERS-CoV infection. J Infect Dis 213:1557–1561
24. Widagdo W, Okba NMA, Richard M, de Meulder D, Bestebroer TM, Lexmond P, Farag E, Al-Hajri M, Stittelaar KJ, de Waal L, van Amerongen G, van den Brand JMA, Haagmans BL, Herfst S (2019) Lack of Middle East respiratory syndrome coronavirus transmission in rabbits. Viruses 11:E381
25. Adney DR, Letko M, Ragan IK, Scott D, van Doremalen N, Bowen RA, Munster VJ (2019) Bactrian camels shed large quantities of Middle East respiratory syndrome coronavirus (MERS-CoV) after experimental infection. Emerg Microbes Infect 8:717–723
26. Adney DR, van Doremalen N, Brown VR, Bushmaker T, Scott D, de Wit E, Bowen RA, Munster VJ (2014) Replication and shedding of MERS-CoV in upper respiratory tract of inoculated dromedary camels. Emerg Infect Dis 20:1999–2005
27. Haagmans BL, van den Brand JM, Raj VS, Volz A, Wohlsein P, Smits SL, Schipper D, Bestebroer TM, Okba N, Fux R, Bensaid A, Solanes Foz D, Kuiken T, Baumgartner W, Segales J, Sutter G, Osterhaus AD (2016) An orthopoxvirus-based vaccine reduces virus excretion after MERS-CoV infection in dromedary camels. Science 351:77–81
28. Munster VJ, de Wit E, Feldmann H (2013) Pneumonia from human coronavirus in a macaque model. N Engl J Med 368:1560–1562
29. Chan JF, Yao Y, Yeung ML, Deng W, Bao L, Jia L, Li F, Xiao C, Gao H, Yu P, Cai JP, Chu H,

Zhou J, Chen H, Qin C, Yuen KY (2015) Treatment with lopinavir/ritonavir or interferon-beta1b improves outcome of MERS-CoV infection in a nonhuman primate model of common marmoset. J Infect Dis 212:1904–1913

30. Falzarano D, de Wit E, Feldmann F, Rasmussen AL, Okumura A, Peng X, Thomas MJ, van Doremalen N, Haddock E, Nagy L, LaCasse R, Liu T, Zhu J, McLellan JS, Scott DP, Katze MG, Feldmann H, Munster VJ (2014) Infection with MERS-CoV causes lethal pneumonia in the common marmoset. PLoS Pathog 10: e1004250
31. Cockrell AS, Johnson JC, Moore IN, Liu DX, Bock KW, Douglas MG, Graham RL, Solomon J, Torzewski L, Bartos C, Hart R, Baric RS, Johnson RF (2018) A spike-modified Middle East respiratory syndrome coronavirus (MERS-CoV) infectious clone elicits mild respiratory disease in infected rhesus macaques. Sci Rep 8:10727
32. Johnson RF, Bagci U, Keith L, Tang X, Mollura DJ, Zeitlin L, Qin J, Huzella L, Bartos CJ, Bohorova N, Bohorov O, Goodman C, Kim DH, Paulty MH, Velasco J, Whaley KJ, Johnson JC, Pettitt J, Ork BL, Solomon J, Oberlander N, Zhu Q, Sun J, Holbrook MR, Olinger GG, Baric RS, Hensley LE, Jahrling PB, Marasco WA (2016) 3B11-N, a monoclonal antibody against MERS-CoV, reduces lung pathology in rhesus monkeys following intratracheal inoculation of MERS-CoV Jordan-n3/2012. Virology 490:49–58
33. Johnson RF, Via LE, Kumar MR, Cornish JP, Yellayi S, Huzella L, Postnikova E, Oberlander N, Bartos C, Ork BL, Mazur S, Allan C, Holbrook MR, Solomon J, Johnson JC, Pickel J, Hensley LE, Jahrling PB (2015) Intratracheal exposure of common marmosets to MERS-CoV Jordan-n3/2012 or MERS-CoV EMC/2012 isolates does not result in lethal disease. Virology 485:422–430
34. de Wit E, Rasmussen AL, Falzarano D, Bushmaker T, Feldmann F, Brining DL, Fischer ER, Martellaro C, Okumura A, Chang J, Scott D, Benecke AG, Katze MG, Feldmann H, Munster VJ (2013) Middle East respiratory syndrome coronavirus (MERS-CoV) causes transient lower respiratory tract infection in rhesus macaques. Proc Natl Acad Sci U S A 110:16598–16603
35. Yao Y, Bao L, Deng W, Xu L, Li F, Lv Q, Yu P, Chen T, Xu Y, Zhu H, Yuan J, Gu S, Wei Q, Chen H, Yuen KY, Qin C (2014) An animal model of MERS produced by infection of rhesus macaques with MERS coronavirus. J Infect Dis 209:236–242
36. Coleman CM, Matthews KL, Goicochea L, Frieman MB (2014) Wild-type and innate immune-deficient mice are not susceptible to the Middle East respiratory syndrome coronavirus. J Gen Virol 95:408–412
37. Zhao J, Li K, Wohlford-Lenane C, Agnihothram SS, Fett C, Zhao J, Gale MJ Jr, Baric RS, Enjuanes L, Gallagher T, McCray PB Jr, Perlman S (2014) Rapid generation of a mouse model for Middle East respiratory syndrome. Proc Natl Acad Sci U S A 111:4970–4975
38. Agrawal AS, Garron T, Tao X, Peng BH, Wakamiya M, Chan TS, Couch RB, Tseng CT (2015) Generation of a transgenic mouse model of Middle East respiratory syndrome coronavirus infection and disease. J Virol 89:3659–3670
39. Li K, Wohlford-Lenane C, Perlman S, Zhao J, Jewell AK, Reznikov LR, Gibson-Corley KN, Meyerholz DK, McCray PB Jr (2016) Middle East Respiratory syndrome coronavirus causes multiple organ damage and lethal disease in mice transgenic for human dipeptidyl peptidase 4. J Infect Dis 213:712–722
40. Zhao G, Jiang Y, Qiu H, Gao T, Zeng Y, Guo Y, Yu H, Li J, Kou Z, Du L, Tan W, Jiang S, Sun S, Zhou Y (2015) Multi-organ damage in human dipeptidyl peptidase 4 transgenic mice infected with Middle East respiratory syndrome-coronavirus. PLoS One 10: e0145561
41. Pascal KE, Coleman CM, Mujica AO, Kamat V, Badithe A, Fairhurst J, Hunt C, Strein J, Berrebi A, Sisk JM, Matthews KL, Babb R, Chen G, Lai KM, Huang TT, Olson W, Yancopoulos GD, Stahl N, Frieman MB, Kyratsous CA (2015) Pre- and postexposure efficacy of fully human antibodies against Spike protein in a novel humanized mouse model of MERS-CoV infection. Proc Natl Acad Sci U S A 112:8738–8743
42. Cockrell AS, Yount BL, Scobey T, Jensen K, Douglas M, Beall A, Tang XC, Marasco WA, Heise MT, Baric RS (2016) A mouse model for MERS coronavirus-induced acute respiratory distress syndrome. Nat Microbiol 2:16226
43. Li K, Wohlford-Lenane CL, Channappanavar R, Park JE, Earnest JT, Bair TB, Bates AM, Brogden KA, Flaherty HA, Gallagher T, Meyerholz DK, Perlman S, McCray PB Jr (2017) Mouse-adapted MERS coronavirus causes lethal lung disease in human DPP4 knockin mice. Proc Natl Acad Sci U S A 114:E3119–E3128
44. Douglas MG, Kocher JF, Scobey T, Baric RS, Cockrell AS (2018) Adaptive evolution influences the infectious dose of MERS-CoV

necessary to achieve severe respiratory disease. Virology 517:98–107

45. Algaissi A, Agrawal AS, Han S, Peng BH, Luo C, Li F, Chan TS, Couch RB, Tseng CK (2019) Elevated human dipeptidyl peptidase 4 expression reduces the susceptibility of hDPP4 transgenic mice to Middle East respiratory syndrome coronavirus infection and disease. J Infect Dis 219:829–835
46. Coleman CM, Sisk JM, Halasz G, Zhong J, Beck SE, Matthews KL, Venkataraman T, Rajagopalan S, Kyratsous CA, Frieman MB (2017) CD8+ T cells and macrophages regulate pathogenesis in a mouse model of Middle East respiratory syndrome. J Virol 91:e01825
47. Fan C, Wu X, Liu Q, Li Q, Liu S, Lu J, Yang Y, Cao Y, Huang W, Liang C, Ying T, Jiang S, Wang Y (2018) A human DPP4-Knockin Mouse's susceptibility to infection by authentic and pseudotyped MERS-CoV. Viruses 10: E448
48. Iwata-Yoshikawa N, Okamura T, Shimizu Y, Kotani O, Sato H, Sekimukai H, Fukushi S, Suzuki T, Sato Y, Takeda M, Tashiro M, Hasegawa H, Nagata N (2019) Acute respiratory infection in human dipeptidyl peptidase 4-transgenic mice infected with Middle East respiratory syndrome coronavirus. J Virol 93: e01818
49. Raj VS, Mou H, Smits SL, Dekkers DH, Muller MA, Dijkman R, Muth D, Demmers JA, Zaki A, Fouchier RA, Thiel V, Drosten C, Rottier PJ, Osterhaus AD, Bosch BJ, Haagmans BL (2013) Dipeptidyl peptidase 4 is a functional receptor for the emerging human coronavirus-EMC. Nature 495:251–254
50. Wang N, Shi X, Jiang L, Zhang S, Wang D, Tong P, Guo D, Fu L, Cui Y, Liu X, Arledge KC, Chen YH, Zhang L, Wang X (2013) Structure of MERS-CoV spike receptor-binding domain complexed with human receptor DPP4. Cell Res 23:986–993
51. Peck KM, Burch CL, Heise MT, Baric RS (2015) Coronavirus host range expansion and Middle East respiratory syndrome coronavirus emergence: biochemical mechanisms and evolutionary perspectives. Annu Rev Virol 2:95–117
52. Barlan A, Zhao J, Sarkar MK, Li K, McCray PB Jr, Perlman S, Gallagher T (2014) Receptor variation and susceptibility to Middle East respiratory syndrome coronavirus infection. J Virol 88:4953–4961
53. Cockrell AS, Peck KM, Yount BL, Agnihothram SS, Scobey T, Curnes NR, Baric RS, Heise MT (2014) Mouse dipeptidyl peptidase 4 is not a functional receptor for Middle East respiratory syndrome coronavirus infection. J Virol 88:5195–5199
54. Peck KM, Cockrell AS, Yount BL, Scobey T, Baric RS, Heise MT (2015) Glycosylation of mouse DPP4 plays a role in inhibiting Middle East respiratory syndrome coronavirus infection. J Virol 89:4696–4699
55. Peck KM, Scobey T, Swanstrom J, Jensen KL, Burch CL, Baric RS, Heise MT (2017) Permissivity of dipeptidyl peptidase 4 orthologs to Middle East respiratory syndrome coronavirus is governed by glycosylation and other complex determinants. J Virol 91:e00534
56. van Doremalen N, Miazgowicz KL, Milne-Price S, Bushmaker T, Robertson S, Scott D, Kinne J, McLellan JS, Zhu J, Munster VJ (2014) Host species restriction of Middle East respiratory syndrome coronavirus through its receptor, dipeptidyl peptidase 4. J Virol 88:9220–9232
57. Klemann C, Wagner L, Stephan M, von Horsten S (2016) Cut to the chase: a review of CD26/dipeptidyl peptidase-4's (DPP4) entanglement in the immune system. Clin Exp Immunol 185:1–21
58. Kameoka J, Tanaka T, Nojima Y, Schlossman SF, Morimoto C (1993) Direct association of adenosine deaminase with a T cell activation antigen, CD26. Science 261:466–469
59. Ohnuma K, Dang NH, Morimoto C (2008) Revisiting an old acquaintance: CD26 and its molecular mechanisms in T cell function. Trends Immunol 29:295–301
60. Weihofen WA, Liu J, Reutter W, Saenger W, Fan H (2004) Crystal structure of CD26/dipeptidyl-peptidase IV in complex with adenosine deaminase reveals a highly amphiphilic interface. J Biol Chem 279:43330–43335
61. Cong L, Ran FA, Cox D, Lin S, Barretto R, Habib N, Hsu PD, Wu X, Jiang W, Marraffini LA, Zhang F (2013) Multiplex genome engineering using CRISPR/Cas systems. Science 339:819–823
62. Mali P, Yang L, Esvelt KM, Aach J, Guell M, DiCarlo JE, Norville JE, Church GM (2013) RNA-guided human genome engineering via Cas9. Science 339:823–826
63. Yang H, Wang H, Shivalila CS, Cheng AW, Shi L, Jaenisch R (2013) One-step generation of mice carrying reporter and conditional alleles by CRISPR/Cas-mediated genome engineering. Cell 154:1370–1379
64. Chefer S, Seidel J, Cockrell AS, Yount B, Solomon J, Hagen KR, Liu DX, Huzella LM, Kumar MR, Postnikova E, Bohannon JK, Lackemeyer MG, Cooper K, Endlich-Frazier A, Sharma H, Thomasson D, Bartos C, Sayre PJ,

Sims A, Dyall J, Holbrook MR, Jahrling PB, Baric RS, Johnson RF (2018) The human sodium iodide symporter as a reporter gene for studying Middle East respiratory syndrome coronavirus pathogenesis. mSphere 3:e00540

65. Leist SR, Baric RS (2018) Giving the genes a shuffle: using natural variation to understand host genetic contributions to viral infections. Trends Genet 34:777–789

66. Menachery VD, Gralinski LE, Mitchell HD, Dinnon KH 3rd, Leist SR, Yount BL Jr, Graham RL, McAnarney ET, Stratton KG, Cockrell AS, Debbink K, Sims AC, Waters KM, Baric RS (2017) Middle East respiratory syndrome coronavirus nonstructural protein 16 is necessary for interferon resistance and viral pathogenesis. mSphere 2:e00346

67. Menachery VD, Mitchell HD, Cockrell AS, Gralinski LE, Yount BL Jr, Graham RL, McAnarney ET, Douglas MG, Scobey T, Beall A, Dinnon K 3rd, Kocher JF, Hale AE, Stratton KG, Waters KM, Baric RS (2017) MERS-CoV accessory ORFs play key role for infection and pathogenesis. MBio 8:e00665

68. Sheahan TP, Sims AC, Graham RL, Menachery VD, Gralinski LE, Case JB, Leist SR, Pyrc K, Feng JY, Trantcheva I, Bannister R, Park Y, Babusis D, Clarke MO, Mackman RL, Spahn JE, Palmiotti CA, Siegel D, Ray AS, Cihlar T, Jordan R, Denison MR, Baric RS (2017) Broad-spectrum antiviral GS-5734 inhibits both epidemic and zoonotic coronaviruses. Sci Transl Med 9:eaal3653

69. Huijbers IJ (2017) Generating genetically modified mice: a decision guide. Methods Mol Biol 1642:1–19

70. Scott GJ, Gruzdev A (2019) Genome editing in mouse embryos with CRISPR/Cas9. Methods Mol Biol 1960:23–40

71. Scobey T, Yount BL, Sims AC, Donaldson EF, Agnihothram SS, Menachery VD, Graham RL, Swanstrom J, Bove PF, Kim JD, Grego S, Randell SH, Baric RS (2013) Reverse genetics with a full-length infectious cDNA of the Middle East respiratory syndrome coronavirus. Proc Natl Acad Sci U S A 110:16157–16162

72. Cockrell AS, Beall A, Yount B, Baric R (2017) Efficient reverse genetic systems for rapid genetic manipulation of emergent and pre-emergent infectious coronaviruses. Methods Mol Biol 1602:59–81

73. Menachery VD, Gralinski LE, Baric RS, Ferris MT (2015) New metrics for evaluating viral respiratory pathogenesis. PLoS One 10: e0131451

Chapter 13

Development of a Mouse-Adapted MERS Coronavirus

Kun Li and Paul B. McCray Jr.

Abstract

First identified in 2012, Middle East respiratory syndrome coronavirus (MERS-CoV) is a novel virus that can cause acute respiratory distress syndrome (ARDS), multiorgan failure, and death, with a case fatality rate of ~35%. An animal model that supports MERS-CoV infection and causes severe lung disease is useful to study pathogenesis and evaluate therapies and vaccines. The murine dipeptidyl peptidase 4 (Dpp4) protein is not a functional receptor for MERS-CoV; thus, mice are resistant to MERS-CoV infection. We generated human DPP4 knock-in (hDPP4 KI) mice by replacing exons 10–12 at the mouse *Dpp4* locus with exons 10–12 from the human DPP4 gene. The resultant human DPP4 KI mice are permissive to MERS-CoV (HCoV-EMC/2012 strain) infection but develop no disease. To generate a mouse model with associated morbidity and mortality from respiratory disease, we serially passaged HCoV-EMC/2012 strain in the lungs of young hDPP4 KI mice. After 30 in vivo passages, an adapted virus clone was isolated and designated $MERS_{MA}$6.1.2. This virus clone produced significantly higher titers than the parental clone in the lungs of hDPP4 KI mice and caused diffuse lung injury and a fatal respiratory infection. In this chapter, we will describe in detail the procedures used to mouse adapt MERS-CoV by serial passage of the virus in lungs. We also describe the methods used to isolate virus clones and characterize virus infection.

Key words Mouse model, Mouse-adapted virus, Serial passage, Virus clone isolation

1 Introduction

Middle East respiratory syndrome (MERS) is a fatal respiratory illness that first appeared on the Saudi Arabian Peninsula in mid-2012. It is caused by a novel betacoronavirus, MERS coronavirus (MERS-CoV) [1]. Shortly after the identification of the virus, its receptor dipeptidyl peptidase 4 (DPP4) was discovered [2]. As of September 2019, the World Health Organization (WHO) has reported 2468 laboratory-confirmed cases of MERS in 27 countries, including 851 associated deaths (fatality rate: ~35%). MERS-CoV does not currently have pandemic potential [3–5]. However, MERS-CoV is still epidemic in the Middle East and remains a cause for significant concern due to the potential spread by global travel as demonstrated by the outbreak in South Korea in 2015 [6–8].

Rahul Vijay (ed.), *MERS Coronavirus: Methods and Protocols*, Methods in Molecular Biology, vol. 2099,
https://doi.org/10.1007/978-1-0716-0211-9_13,

To date, there are only two published reports on autopsy findings from subjects who died from MERS [9, 10] and our understanding of MERS-CoV pathogenesis in humans is still limited. Animal models are useful for the study of viral diseases and play an important role in the investigation of pathogenesis and evaluation of antiviral therapies and vaccines. An ideal animal model should be permissive to the viral infection and develop disease and pathology with similarities to that observed in humans. MERS-CoV infection has been evaluated in two nonhuman primate (NHP) models, the rhesus macaque and common marmoset [11–15]. Both species are susceptible to MERS-CoV infection. However, MERS-CoV caused only a transient lower respiratory tract infection without mortality in rhesus macaques [11–13]. In common marmosets, the consequences of MERS-CoV infection are controversial. Falzarano et al. reported the common marmoset reproduced several features of MERS-CoV infection in humans including progressive severe pneumonia [15], while another group observed only mild to moderate nonlethal respiratory disease following MERS-CoV infection [14]. Thus, the common marmoset is potentially a model to study pathogenesis and evaluate antiviral therapies and vaccines. However, NHPs are expensive, their availability limited, and their use may raise ethical concerns. In contrast, small animal models provide advantages over NHPs, including reduced cost, availability in large numbers, ease of handling, and species-specific reagents, especially for the studies of highly pathogenic viruses like MERS-CoV in the Biosafety Level 3 (BSL-3) laboratory. Unfortunately, common small laboratory animals like mice [16, 17], ferrets [18], guinea pigs [19], and hamsters [20] are not susceptible to MERS-CoV infection because their homologous DPP4 cannot be bound and utilized by MERS-CoV as a host receptor for entry [21, 22]. Haagmans et al. detected MERS-CoV RNA in the respiratory tract of New Zealand white rabbits following inoculation but found no clinical signs of disease [23].

Several strategies have been used to overcome this receptor incompatibility and develop mouse models of MERS-CoV infection. In 2014, we developed the first mouse model of MERS-CoV infection [24]. We delivered a recombinant adenovirus 5 encoding human DPP4 (hDPP4) to the lungs of mice. Transient expression of hDPP4 by adenovirus transduction made the mice temporarily permissive for MERS-CoV infection, but animals developed only mild lung disease. Generation of transgenic mice expressing a virus receptor is a common strategy to make mice permissive to infection. Several groups in addition to ours developed mice with transgenic expression of hDPP4 using different promoters [25–28]. The disease severity following MERS-CoV infection in transgenic mice correlated with the cellular distribution and expression level of hDPP4. In transgenic mice expressing hDPP4 driven by the cytokeratin 18 promoter [26] or a ubiquitous promoter [25, 27, 28],

MERS-CoV replicates and causes respiratory disease and mortality. However, the lethality was found to be secondary to overwhelming central nervous system (CNS) disease or multiorgan damage. This was also observed in mice transgenic for human ACE2 driven by the cytokeratin 18 promoter and infected with SARS-CoV [29]. In transgenic mice expressing hDPP4 under the human surfactant protein C (SPC) promoter, which restricts expression to bronchiolar and alveolar epithelia, MERS-CoV infection caused only mild disease [26]. Thus, these transgenic mice do not reproduce a severe lung disease phenotype that resembles MERS.

Alternative strategies for the creation of mouse models of MERS-CoV infection are generation of DPP4 humanized mice and adaptation of the virus to the animals. Pascal et al. reported a model in which all of the mouse *Dpp4* exons had been humanized and also generated humanized monoclonal antibodies against the MERS-CoV S protein using a novel strategy [30]. MERS-CoV infection in this model caused pulmonary edema, vascular cuffing, and alveolar septal thickening with an associated ~20% weight loss, necessitating euthanasia [31]. Another MERS mouse model was engineered by changing two amino acids in the mouse *Dpp4* locus using CRISPR-Cas9 technology [32]. This model supported MERS-CoV replication without severe disease. Similarly, our human DPP4 knock-in mouse model supported MERS-CoV replication but did not lead to a severe lung disease phenotype [33]. Two mouse-adapted (MA) strains of MERS-CoV were subsequently developed independently by serial passage of the HCoV-EMC/2012 strain [1] in the lungs of the two humanized mouse models [32, 33]. The resultant MERS-15 and $MERS_{MA}6.1.2$ mouse-adapted MERS-CoV strains replicated to high titers in the lungs of the CRISPR-Cas9 genetically engineered mouse model and the hDPP4 knock-in mouse model, respectively. The respiratory disease that developed in both mouse models and the associated mortality shared similarities with severe cases of MERS [32, 33]. Mouse adaptation was also successfully used to generate several SARS-CoV strains capable of modeling severe SARS-CoV lung disease in mice [34–36]. Thus, the adaption of the virus to enhance virulence in the mouse is a very useful approach to generate mouse models for coronavirus-associated lung disease.

2 Materials

These materials can be altered to fit the requirements for other viruses of interest.

1. HCoV-EMC/2012 strain.
2. hDPP4 knock-in mouse (C57BL/6 strain with mouse *Dpp4* exons 10–12 replaced with the human codons).

3. Insulated foam box filled with ice.
4. 2.0 mL sterile plastic screw-top tubes with O-ring caps.
5. Sterile DMEM (serum free).
6. 1 mL insulin syringe fitted with a 28 gauge (g) × ½ inch needle
7. Ketamine/xylazine (87.5 mg/kg ketamine/12.5 mg/kg xylazine).
8. Pipettes and filtered pipette tips: 10 μL, 20 μL, 200 μL, 1 mL.
9. 1000 mL polypropylene beaker
10. A scale with sensitivity in the 10–50 g range for weighing mice.
11. A flat-bottomed container with sides that fits on the weighing platform of the scale and can prevent mice from escaping (e.g., a 400 mL disposable beaker).
12. 30 cm long metal straight forceps to transfer mice from the cage
13. Transparent vacuum desiccator.
14. Polystyrene foam surface (e.g., a polystyrene box top).
15. Disposable absorbent bench underpads.
16. Precision glide needles 25 g × 5/8 inch.
17. Spray bottle filled with 70% ethanol.
18. Spray bottle filled with Virex Plus.
19. Surgical tools (e.g., scissors, straight forceps, curved forceps, single edge razor blades, curved serrated forceps).
20. 1× Dulbecco's phosphate buffered saline (DPBS)
21. 10 mL syringes.
22. Small polystyrene weighing dish.
23. Paper towels.
24. 50 mL disposable tissue grinder.
25. TRIzol reagent.
26. Vero81 cells.
27. Huh7 cells.
28. Sterile 6-well plates and 12-well plates.
29. 37 °C incubator containing 5% CO_2.
30. D10: 1× DMEM, 10% FBS, and 1% penicillin/streptomycin (PS).
31. D2: 1× DMEM, 2% FBS, and 1% PS.
32. Overlay media: 2× DMEM, 4% FBS, and 2% PS.
33. 2% Low melting point agarose in ddH_2O.
34. 10 mL stripettes.
35. Electronic pipette controller.
36. 1 ml graduated transfer pipettes.

37. Water bath.
38. 15 mL sterile conical tube.
39. Dry ice.

3 Methods

All procedures are performed under BSL-3 laboratory conditions and must follow the standard operating protocol of a BSL-3 facility and regulatory agencies.

All manipulations of infectious specimens, samples, and mice must be performed within a biosafety cabinet or within a contained device such as a centrifuge.

All tissue culture media and waste must be bleached and autoclaved prior to disposal.

All instruments must be disinfected with Virex plus or 10% bleach.

3.1 Intranasal Infection

1. Rapidly thaw aliquots of virus (*see* **Note 1**). Place the virus stock aliquots on ice immediately after thawed.
2. Dilute the stock virus in 2.0 mL sterile screw-top tubes with O-ring cap with ice-cold serum-free DMEM to desired inoculum and keep the diluted virus on ice during mouse infection (*see* **Note 2**).
3. Anesthetize mice by intraperitoneal injection of ketamine/xylazine (87.5 mg/kg ketamine/12.5 mg/kg xylazine) (*see* **Note 3**).
4. Hold the mouse in a vertical position with the nose pointed upward (*see* **Note 4**). Carefully pipette 50 μL of diluted virus onto the nostrils drop by drop, carefully matching the rate at which the mouse inhales (*see* **Note 5**).
5. Place inoculated mouse in a 1000 mL polypropylene beaker. Repeat until all mice are infected. Transfer mice back to their cage and monitor until they regain consciousness (*see* **Note 6**).

3.2 Evaluation of Virulence by Weight Loss and Survival

1. Check mice and monitor weight every day throughout the duration of virus challenge (*see* **Note 7**). To evaluate weight, lift the mouse by the tail using the long forceps and place the mouse into the container on the scale.
2. Record weight when the scale reading is stable.
3. Transfer the mouse into a 1000 mL polypropylene beaker. Repeat until all mice are weighed. Transfer mice from beaker back to the cage.

4. Calculate the percentage of weight loss normalized to the starting weight. Euthanize mouse if the weight loss is $\geq$30% of starting weight (*see* **Note 8**).

3.3 Harvest Virus from the Lungs and Other Organs

1. At indicated days post infection, mice should be anesthetized by intraperitoneal injection of ketamine/xylazine.
2. Once the right plane of anesthesia is attained immediately, place the mouse ventral side up on a polystyrene foam surface covered with absorbent bench underpad (*see* **Note 9**). Immobilize the mouse by pinning limbs to the foam with 25 g × 5/8 inch needles.
3. Wet the ventral side of the mouse with 70% ethanol. Pinch the fur/skin near the urethral opening with forceps and pull slightly upward. Make a midline incision through the fur/skin from urethral opening up to the mandible with surgical scissors. Using forceps, peel the fur/skin transversely to both sides along the incision to separate fur/skin from underlying muscles.
4. Lift abdominal muscle and incise through midline up to the base of the thorax. Then make transverse incisions to open the abdominal cavity.
5. Insert surgical scissors under the sternum and cut the diaphragm following the costal arch. Remove the rib cage using scissors and forceps, exposing the lungs and heart.
6. Fill a 10 mL syringe with cold sterile DPBS using a 25 g × 5/8 inch needle. Insert the needle into the apex of the left ventricle and make a small incision in the right atrium. Slowly perfuse $\geq$5 mL cold sterile DPBS into the left ventricle. Next, insert the needle into the apex of the right ventricle and perfuse $\geq$5 mL cold sterile DPBS (*see* **Note 10**).
7. Remove the lungs and heart from the thoracic cavity. Remove the liver, kidney, spleen, and small intestine. Place organs in a small polystyrene weighing dish. Remove remaining connective tissue.
8. To harvest the brain, turn the mouse over and wet the fur of the head with 70% ethanol. Grasp the ears with forceps and cut off the skin and fur to expose the skull. Remove remaining skin at the base of the neck to further expose the skull. Immobilize the mouse and make an incision along the sagittal suture of the skull using a single edge razor blade (*see* **Note 11**). Wedge one prong of the curved serrated forceps into the now opened sagittal suture. Slowly pry up the skull, grasp the piece of skull with forceps and peel outward to remove. Repeat on the other side. Use curved forceps to lift the brain from the skull. Place brain tissue in the small polystyrene weighing dish.

9. Carefully place each organ into a 50 mL disposable tissue grinder filled with 2 mL sterile cold DPBS for homogenization or into 2 mL of TRIzol for RNA extraction (*see* **Note 12**).
10. Grind the lung tissue and transfer the homogenates or RNA samples into 2.0 mL sterile screw-top tubes with O-ring cap (*see* **Note 13**). Store samples at −80 ° C. Thaw and spin down the cell debris in lung homogenates before use.

3.4 Adaptation of the Virus to Mouse by Serial Passage in the Lung

1. To mouse adapt the virus, infect two mice intranasally with 10^5 pfu/mouse HCoV-EMC/2012 strain (*see* **Note 14**).
2. Two days after infection, prepare lung homogenates from the two mice.
3. Combine 100 μL of lung homogenate from each mouse in a 2.0 mL sterile screw-top tube with O-ring cap on ice. Then inoculate two new mice intranasally with 50 μL/mouse of the mixed lung tissue homogenates.
4. Repeat this process (**steps 2–3**). The virulence of the virus should be evaluated in groups of mice by weight loss and survival after every 5–10 in vivo passages.

3.5 Isolation of Viral Clones

After the virulence of the virus has been significantly enhanced, single plaques of the adapted virus should be purified and evaluated.

1. Plate Vero81 cells in D10 media in 6-well plates one day before infection.
2. Rapidly thaw lung homogenates from the selected passage. Mix the lung homogenates from two mice in a 2.0 mL sterile screw-top tube with O-ring cap on ice.
3. Serially dilute the mixed lung homogenates tenfold in 2.0 mL sterile screw-top tubes with O-ring caps, using ice-cold serum-free DMEM (*see* **Note 15**). Keep the dilutions on ice.
4. Remove the medium from each well and add diluted samples (in a volume of 400 μL) to each well.
5. Place the plates in the 37 °C incubator for 1 h and rotate gently every 15 min.
6. Melt 2% low melting point agarose and maintain in a 65 °C water bath.
7. Mix overlay media and 2% low melting point agarose at a volume ratio of 1:1. Rotate the tube several times to fully mix. Overlay cells with 1.5 mL of mixed media using 10 mL stripettes.
8. Let the plates sit in the hood for ~5 min at RT or until the agarose overlay turns solid. Add 0.5 mL D2 medium on the top of solidified agarose.

9. Place the plates in the 37 °C incubator.
10. After 3 days, plaques should be visible. Remove the liquid on the top of the agarose. Circle the visible plaques on the underside of the plates using a permanent marker (*see* **Note 16**).
11. Vertically penetrate the agarose and pipette the circled plaque several times with a 1 mL graduated transfer pipette. The agarose above the plaque will be pulled into the pipette.
12. Transfer the agarose above the plaque into a 15 mL conical tube filled with 500 μL DMEM by pipetting up and down several times.
13. Transfer the 500 μL DMEM containing the agarose into a 2.0 mL sterile screw-top tube with O-ring cap.
14. Repeat the procedure and pick six single plaques. Store the tubes at −80 ° C.
15. Thaw the tubes containing the isolated plaques on ice.
16. Propagate viruses from the isolated plaques in Huh7 cells (*see* **Note 17**). To do this, plate Huh7 cells in D10 media in a 6-well plate one day before infection. Remove medium and add 500 μL of DMEM containing the agarose into the well. Place in the 37 °C incubator for 1 h Remove the 500 μL DMEM and add 2 mL D2 media.
17. Place the plates in the 37 °C incubator.
18. At 1 day post infection, freeze cells by putting the plate on dry ice and then thawing the cell. Harvest cell lysate and store at −80 ° C.
19. Titrate the serially passaged viruses by plaque assay on Vero81 cells. Briefly, Vero81 cells are plated in a 12-well plate. On the next day, thaw the virus and serially dilute the virus by tenfold for at least three dilutions with serum-free medium. Remove medium from the wells (Vero81 plate) and add 250 μL of diluted virus into each well of the plate. Incubate the plate at 37 °C incubator for 1 h and with gentle rotations of the plate every 15 min to prevent drying of cells. In the meantime, melt 2% low melting point agarose and maintain it in a 65 °C water bath. After 1 h of incubation, remove the 250 μL inoculum and overlay cells with 1 mL of 1:1 mix of 2× DMEM and 2% low melting point agarose. After the agar overlay turns solid, add two drops of 2% FBS DMEM medium on the top of the solid agar by 10 mL stripettes. The plate should be incubated at 37 °C and 5% CO_2 for about 3 days. After 3 days, fix the wells by adding 25% formaldehyde to completely fill the wells for 20 min. Then aspirate the liquid from each well and remove the agarose. Stain the wells with 0.1% crystal violet to delineate plaques. Remove the crystal violet, rinse wells with DPBS, and count plaques in the hood.

20. Evaluate the virulence of the plaque-isolated virus by monitoring weight loss and survival in mice after intranasal infection with 10^5 pfu/mouse as mentioned above.
21. Characterize virus distribution by titrating virus load in various organ homogenates and virus genomic RNA abundance in tissues using qRT-PCR.
22. Characterize lung histology after infection.
23. Use next-generation sequencing to identify changes in the MERS-CoV genome sequence following serial passage [33].

4 Notes

1. To rapidly thaw the frozen virus, immerse the vial in a plastic beaker filled with water at room temperature within the biosafety cabinet. The rate of thawing will depend on the volume (approximately 2 min or until ice crystals melt).
2. Always keep freshly thawed virus aliquots or dilutions on ice. This makes the virus more stable and leads to more reproducible results.
3. Check that mice are fully anesthetized by observing the pedal reflex. If mice are not fully anesthetized, they may move excessively or sneeze when virus is applied.
4. While holding mice, make sure no pressure is applied over the throat area to avoid interference with respiration.
5. Do not place the pipette tip inside the nostril. If a mouse does not inhale the droplet, stop pipetting, and wait until the mouse inhales what is already applied.
6. Monitor the health of mice before returning to the cage.
7. Record weights from the day of infection until recovery.
8. Animal euthanasia should follow the institution's Animal Care and Use guidelines. There may be differences in institutional requirements regarding when euthanasia is required based on weight loss.
9. Use the lid of an insulated foam shipping box. Cut absorbent bench underpad (42 × 58 cm) into small pieces that fit the lid.
10. Carefully keep the tip of the needle in the lumen of the ventricle. Be sure to puncture the right atrium as this will help to drain blood during the perfusion. The lung will turn white after perfusion.
11. Make sure that the incision does not exceed the thickness of the skull; avoid cutting into the brain tissue.
12. The individual organs can be divided into pieces and transferred to grinders with PBS or TRIzol separately.

13. Lung homogenates are immediately transferred to tubes on ice. RNA samples are transferred to a tube and incubated for at least 15 min at room temperature per our BSL-3 specific standard operating procedures.
14. At 2 days post infection with 10^5 pfu/mouse, HCoV-EMC/2012 strain replicates in the lung of the hDPP4 KI mice to a titer of around 4×10^6 pfu/mL, which equals 2×10^5 pfu in 50 μL. We chose the 10^5 pfu/mouse inoculum to begin because we wanted to use a similar dose range during in vivo serial passage while skipping the titration step.
15. Titrate the homogenates of the passage of interest and select a virus dilution that produces ≤ 10 plaques in a well. This allows identification of individual clear single plaques.
16. Only circle the unambiguous clear single plaques.
17. Compared to Vero 81 cells, we found that the MERS-CoV RNA genome is more stable when propagated in Huh7 cells (less likely to introduce genomic deletions, insertions, or point mutations).

Acknowledgements

We thank Chris Wohlford-Lenane and Jennifer Bartlett for careful review of the manuscript. This work is supported by the National Institutes of Health P01 AI060699. We also acknowledge the support of the Cell Morphology Core and Pathology Core, partially supported by the Center for Gene Therapy for Cystic Fibrosis (National Institutes of Health P30 DK-54759) and the Cystic Fibrosis Foundation. P.B.M. is supported by the Roy J. Carver Charitable Trust.

References

1. Zaki AM et al (2012) Isolation of a novel coronavirus from a man with pneumonia in Saudi Arabia. N Engl J Med 367(19):1814–1820
2. Raj VS et al (2013) Dipeptidyl peptidase 4 is a functional receptor for the emerging human coronavirus-EMC. Nature 495 (7440):251–254
3. Breban R, Riou J, Fontanet A (2013) Interhuman transmissibility of Middle East respiratory syndrome coronavirus: estimation of pandemic risk. Lancet 382(9893):694–699
4. Perlman S, McCray PB Jr (2013) Person-to-person spread of the MERS coronavirus-an evolving picture. N Engl J Med 369 (5):466–467
5. Kucharski AJ, Althaus CL (2015) The role of superspreading in Middle East respiratory syndrome coronavirus (MERS-CoV) transmission. Euro Surveill 20(25):14–18
6. Hui DS, Perlman S, Zumla A (2015) Spread of MERS to South Korea and China. Lancet Respir Med 3(7):509–510
7. Zumla A et al (2016) Infectious diseases epidemic threats and mass gatherings: refocusing global attention on the continuing spread of the Middle East respiratory syndrome coronavirus (MERS-CoV). BMC Med 14(1):132
8. Chen X et al (2017) Comparative epidemiology of Middle East respiratory syndrome coronavirus (MERS-CoV) in Saudi Arabia and South Korea. Emerg Microbes Infect 6(6):e51
9. Ng DL et al (2016) Clinicopathologic, immunohistochemical, and ultrastructural findings of a fatal case of Middle East respiratory

syndrome coronavirus infection in the United Arab Emirates, April 2014. Am J Pathol 186 (3):652–658
10. Alsaad KO et al (2018) Histopathology of Middle East respiratory syndrome coronovirus (MERS-CoV) infection - clinicopathological and ultrastructural study. Histopathology 72 (3):516–524
11. de Wit E et al (2013) Middle East respiratory syndrome coronavirus (MERS-CoV) causes transient lower respiratory tract infection in rhesus macaques. Proc Natl Acad Sci U S A 110(41):16598–16603
12. Yao Y et al (2014) An animal model of MERS produced by infection of rhesus macaques with MERS coronavirus. J Infect Dis 209 (2):236–242
13. Prescott J et al (2018) Pathogenicity and viral shedding of MERS-CoV in immunocompromised rhesus macaques. Front Immunol 9:205
14. Johnson RF et al (2015) Intratracheal exposure of common marmosets to MERS-CoV Jordan-n3/2012 or MERS-CoV EMC/2012 isolates does not result in lethal disease. Virology 485:422–430
15. Falzarano D et al (2014) Infection with MERS-CoV causes lethal pneumonia in the common marmoset. PLoS Pathog 10(8):e1004250
16. Cockrell AS et al (2014) Mouse dipeptidyl peptidase 4 is not a functional receptor for Middle East respiratory syndrome coronavirus infection. J Virol 88(9):5195–5199
17. Coleman CM et al (2014) Wild-type and innate immune-deficient mice are not susceptible to the Middle East respiratory syndrome coronavirus. J Gen Virol 95(Pt 2):408–412
18. Raj VS et al (2014) Adenosine deaminase acts as a natural antagonist for dipeptidyl peptidase 4-mediated entry of the Middle East respiratory syndrome coronavirus. J Virol 88 (3):1834–1838
19. de Wit E et al (2017) Domestic pig unlikely reservoir for MERS-CoV. Emerg Infect Dis 23 (6):985–988
20. de Wit E et al (2013) The Middle East respiratory syndrome coronavirus (MERS-CoV) does not replicate in Syrian hamsters. PLoS One 8 (7):e69127
21. van Doremalen N et al (2014) Host species restriction of Middle East respiratory syndrome coronavirus through its receptor, dipeptidyl peptidase 4. J Virol 88(16):9220–9232
22. Barlan A et al (2014) Receptor variation and susceptibility to Middle East respiratory syndrome coronavirus infection. J Virol 88 (9):4953–4961
23. Haagmans BL et al (2015) Asymptomatic Middle East respiratory syndrome coronavirus infection in rabbits. J Virol 89(11):6131–6135
24. Zhao J et al (2014) Rapid generation of a mouse model for Middle East respiratory syndrome. Proc Natl Acad Sci U S A 111 (13):4970–4975
25. Agrawal AS et al (2015) Generation of a transgenic mouse model of Middle East respiratory syndrome coronavirus infection and disease. J Virol 89(7):3659–3670
26. Li K et al (2016) Middle East respiratory syndrome coronavirus causes multiple organ damage and lethal disease in mice transgenic for human dipeptidyl peptidase 4. J Infect Dis 213(5):712–722
27. Zhao G et al (2015) Multi-organ damage in human dipeptidyl peptidase 4 transgenic mice infected with Middle East respiratory syndrome-coronavirus. PLoS One 10(12): e0145561
28. Fan C et al (2018) A human DPP4-knockin mouse's susceptibility to infection by authentic and pseudotyped MERS-CoV. Viruses 10(9): E448
29. McCray PB Jr et al (2007) Lethal infection of K18-hACE2 mice infected with severe acute respiratory syndrome coronavirus. J Virol 81 (2):813–821
30. Pascal KE et al (2015) Pre- and postexposure efficacy of fully human antibodies against Spike protein in a novel humanized mouse model of MERS-CoV infection. Proc Natl Acad Sci U S A 112(28):8738–8743
31. Coleman CM et al (2017) CD8+ T cells and macrophages regulate pathogenesis in a mouse model of Middle East respiratory syndrome. J Virol 91(1):e01825
32. Cockrell AS et al (2016) A mouse model for MERS coronavirus-induced acute respiratory distress syndrome. Nat Microbiol 2:16226
33. Li K et al (2017) Mouse-adapted MERS coronavirus causes lethal lung disease in human DPP4 knockin mice. Proc Natl Acad Sci U S A 114(15):E3119–E3128
34. Roberts A et al (2007) A mouse-adapted SARS-coronavirus causes disease and mortality in BALB/c mice. PLoS Pathog 3(1):e5
35. Day CW et al (2009) A new mouse-adapted strain of SARS-CoV as a lethal model for evaluating antiviral agents in vitro and in vivo. Virology 395(2):210–222
36. Frieman M et al (2012) Molecular determinants of severe acute respiratory syndrome coronavirus pathogenesis and virulence in young and aged mouse models of human disease. J Virol 86(2):884–897

Chapter 14

Metabolite, Protein, and Lipid Extraction (MPLEx): A Method that Simultaneously Inactivates Middle East Respiratory Syndrome Coronavirus and Allows Analysis of Multiple Host Cell Components Following Infection

Carrie D. Nicora, Amy C. Sims, Kent J. Bloodsworth, Young-Mo Kim, Ronald J. Moore, Jennifer E. Kyle, Ernesto S. Nakayasu, and Thomas O. Metz

Abstract

Mass spectrometry (MS)-based, integrated proteomics, metabolomics, and lipidomics (collectively, multi-omics) studies provide a very detailed snapshot of virus-induced changes to the host following infection and can lead to the identification of novel prophylactic and therapeutic targets for preventing or lessening disease severity. Multi-omics studies with Middle East respiratory syndrome coronavirus (MERS-CoV) are challenging as the requirements of biosafety level 3 containment limit the numbers of samples that can be safely managed. To address these issues, the multi-omics sample preparation technique MPLEx (metabolite, protein, and lipid extraction) was developed to partition a single sample into three distinct parts (metabolites, proteins, and lipids) for multi-omics analysis, while simultaneously inactivating MERS-CoV by solubilizing and disrupting the viral envelope and denaturing viral proteins. Here we describe the MPLEx protocol, highlight the step of inactivation, and describe the details of downstream processing, instrumental analysis of the three separate analytes, and their subsequent informatics pipelines.

Key words MPLEx, Metabolomics, Proteomics, Lipidomics, MERS-CoV, Virus-host interactions, Mass spectrometry (MS), Virus inactivation

1 Introduction

Middle East respiratory syndrome coronavirus (MERS-CoV) research is of global interest due to the high case fatality rate, narrowly defined epidemiology, and spread of the virus to 27 countries to date. The World Health Organization (WHO) recognizes the urgent need for effective public health countermeasures due to the ongoing epidemic, and MERS-CoV is on the WHO list of priority pathogens to highlight the critical need for the development of diagnostic products and prophylactic and

Rahul Vijay (ed.), *MERS Coronavirus: Methods and Protocols*, Methods in Molecular Biology, vol. 2099,
https://doi.org/10.1007/978-1-0716-0211-9_14,

therapeutic treatment options [1]. The National Institute of Allergy and Infectious Diseases Systems Biology for Infectious Disease Research Program was established to support research focusing on multi-omics approaches and dataset integration to develop and validate predictive models of infectious disease initiation, progression, and outcomes [2]. Thus, there is great interest and need for improved understanding of the pathogenesis of MERS-CoV; however, challenges remain in effective study of this pathogen.

For example, samples from MERS-CoV infected patients are virtually unavailable for analysis, and animal models that recapitulate disease phenotypes seen in humans have only recently been generated, both of which have drastically slowed our progress toward understanding MERS-CoV pathogenesis in the host [3]. Systems biology studies offer a way to capture big picture snapshots of individual cellular components (proteins, lipids, metabolites) that are modulated over the course of infection to develop a better understanding of pathogen-host interactions [4–13].

Proteins are the major effectors of cellular pathways and represent the dynamic expression of information encoded within the genome during infection. Protein driven cellular responses following infection can favor either viral clearance or spread; therefore, taking snapshots of total proteins isolated from infected cells over the course of infection can provide insights into their underlying molecular mechanisms of pathogenicity, and potentially even single out targets for pharmacological intervention [14].

Metabolites are biomolecules required for cellular metabolism and can either be intermediates produced during cellular metabolic processes or end products of cellular pathways. They represent the level of homeostasis of cellular activities in a host [15, 16]. Importantly, certain metabolites play key roles during the cellular responses to various viral infections such as signaling, initiating or resolving inflammation, or other immune related responses [17]. Therefore, metabolite levels can be profiled between healthy and disease states to not only understand the triggers of change but to also discover possible biomarkers in early disease stages.

Lipids have key functions in signaling pathways, energy storage, and the structural integrity of cell membranes. They also function in host-pathogen interactions and immunomodulation since they act in first-line recognition and host cell signaling during pathogen docking, invasion, and intracellular trafficking [18]. Lipid metabolism and cellular lipids are greatly affected by virus infections by inducing major lipid modifications within host cells through the production of convoluted membranes and double membrane vesicles (DMVs) [19–22]. Virus-induced production of membrane networks and organelles is a common occurrence among all positive sense RNA viruses [23, 24]. The roles of these virus-induced DMVs are not fully understood; however, evidence suggests some viruses may use them for replication, to conceal viral RNA from host antiviral response, or they may have roles in autophagy as autophagosomes [25].

While it's clear that metabolites, proteins, and lipids play an important role in fully characterizing the MERS-CoV infection, the proteomic, metabolomic, and lipidomic sample manipulation of MERS-CoV outside of appropriate biosafety level (BSL) containment laboratories can take place only subsequent to pathogen inactivation. Here, we describe the MPLEx (metabolite, protein, and lipid extraction) protocol for the extraction of protein, metabolites, and lipids from a single sample that simultaneously inactivates the MERS-CoV virus [5, 6, 26]. Each analyte can then be analyzed by the respective mass spectrometry (MS)-based omics pipeline.

To illustrate the effectiveness of this protocol, Nakayasu et al. performed an integrative multi-omics study using a human lung epithelial cell line infected with MERS-CoV, which showed the impact of the viral infection on the host glycolytic pathway, different host metabolic pathways, and also global changes in lipid profiles induced by infection [5]. To illustrate the effectiveness of MPLEx on pathogen inactivation, Burnum-Johnson et al. showed complete inactivation of both bacterial and viral pathogens with exposed lipid envelopes, including MERS-CoV [6].

The MPLEx method is a simple yet powerful protocol that can be applied for integrative multi-omic measurements while concurrently inactivating MERS-CoV (or other enveloped viruses). The multiple analyte samples obtained from MPLEx can be used across various instrument and data analysis platforms. Here we describe the pipeline as implemented at Pacific Northwest National Laboratory.

2 Materials

2.1 Extraction Chemicals, Quality Controls, and Mobile Phases

1. 18.2 MΩ cm water (*see* **Note 1**).
2. Ammonium bicarbonate buffer (NH_4HCO_3): 150 mM, pH 8.0, ice cold.
3. Rapid Quench Buffer: 60% methanol, 0.85% ammonium bicarbonate in sterile water. Store at −80 °C and stand in biological safety cabinet to warm up during processing.
4. MPLEx Solution: two volumes chloroform to one volume methanol (2:1, v/v). Store at −80 °C and stand in biological safety cabinet to warm up to −20 °C during processing (*see* **Note 2**).
5. High-performance liquid chromatography (HPLC) solvent A: 10 mM ammonium formate, pH 10.0.

6. HPLC solvent B: 10:90 10 mM ammonium formate, pH 10.0: acetonitrile (ACN).
7. Proteomics MS mobile phase (MP) A: 0.1% formic acid in water.
8. Proteomics MS MP B: 0.1% formic acid in ACN.
9. Lipidomics MS MP A: ACN:H_2O (40:60) in 10 mM ammonium acetate.
10. Lipidomics MS MP B: ACN:isopropyl alcohol (10:90) in 10 mM ammonium acetate.
11. Metabolomics MS MP: Helium gas.
12. Metabolomics retention time alignment compounds: FAMEs (fatty acid methyl esters) (Agilent, Santa Clara, CA) [27].
13. Lipid quality control: Bovine Brain Total Lipid Extract (BTLE) (Avanti Polar Lipids, Inc., Alabaster, AL).

2.2 Metabolite Derivatization Chemicals

1. Anhydrous pyridine (99.8% purity) (*see* **Note 3**).
2. *O*-Methylhydroxylamine hydrochloride (>98% purity).
3. Silylation Reagent: *N*-Methyl-*N*-(trimethylsilyl) trifluoroacetamide with 1% trimethylchlorosilane (MSTFA with 1% of TMCS) (GC grade) (*see* **Note 4**).

2.3 Tryptic Digestion, Solid Phase Extraction, Tandem Mass Tag (TMT) Isobaric Labeling, and High-Performance Liquid Chromatography (HPLC) Fractionation Chemicals

1. NH_4HCO_3 buffer: 100 mM, pH ~8.4 (*see* **Note 5**).
2. Bicinchoninic acid (BCA) assay reagent kit (Thermo Scientific, Waltham, MA).
3. Urea protein denaturing solution: 8 M in 100 mM NH_4HCO_3, pH ~8.0.
4. Protein-reducing solution: 0.5 M dithiothreitol (DTT) in 100 mM NH_4HCO_3, pH 8.4.
5. Protein alkylation solution: 0.4 M iodoacetamide (IAM) in 100 mM NH_4HCO_3, pH 8.4.
6. 1 M $CaCl_2$ in 100 mM NH_4HCO_3, pH 8.4.
7. Sequencing-grade modified trypsin (*see* **Note 6**).
8. C-18 Solid Phase Extraction (SPE) column conditioning 1: methanol.
9. C-18 SPE column conditioning 2: 0.1% trifluoroacetic acid (TFA) in H_2O.
10. C-18 SPE column washing: 95:5H_2O:ACN, 0.1% TFA.
11. C-18 SPE column elution: 80:20 ACN:H_2O, 0.1% TFA.
12. Amine-reactive Thermo Scientific Tandem Mass Tag (TMT) Isobaric Mass Tagging Kits (Thermo Scientific, Rockford, IL).
13. TMT resuspension reagent: Anhydrous ACN.

14. TMT dissolution buffer: 50 mM triethylammonium bicarbonate (TEAB), pH 8.5.
15. TMT reaction quenching buffer: 5% hydroxylamine.

2.4 Personal Protective Equipment (PPE) and Lab Supplies

1. Milli-Q water purification system.
2. Vortex.
3. Refrigerated centrifuge.
4. Thermomixer with Thermotop.
5. Magnetic stir plate and bars.
6. pH paper strips, pH range 0–14.
7. Breathe Easier plate membranes.
8. −20 °C freezer.
9. −80 °C freezer.
10. Conical glass vials for lipids.
11. Glass vial lids without pre-slit septa.
12. Glass MS sample vial and inserts for proteomics.
13. Pipette set (2–1000 μL).
14. Kimwipes.
15. Chloroform-resistant polypropylene pipette tips (*see* **Note 7**).
16. Chloroform compatible 1.7 mL or 2 mL Sorenson MμlTI™ SafeSeal™ Microcentrifuge Tubes (*see* **Note 8**).
17. Hamilton Glass Syringes, 5 mL, 50 μL, and 250 μL.
18. Bath Sonicator.
19. C-18 solid phase extraction columns 50 mg/1 mL.
20. Liquid nitrogen.
21. Speedvac Concentrator (for drying in vacuo).
22. Biocontainment rotor Eppendorf centrifuge.
23. Serological pipette.
24. Gilson pipetman and non-aerosol tips (Gilson, Middleton, WI).
25. 3 M Powered air-purifying respirator.
26. 3 M Versaflo hood with shroud.
27. 3 M Heavy-duty rubber breathing tube.
28. 3 M Belt for PAPR.
29. 3 M Cartridge filer.
30. 3 M Filter cover.
31. 3 M Battery.
32. 3 M Battery charger.
33. 3 M Airflow indicator.

34. DuPont Tyvek IsoClean Coveralls.
35. DuPont ProClean Boot Covers.
36. Kappler ProVent 7000 Isolation Gowns.
37. Cardinal Health Esteem Nitrile Gloves with Neu-Thera.
38. 70% Ethanol (surface decontamination).
39. CiDecon™ Disinfectant—Concentrated phenolic disinfectant.
40. Autoclave (Steris).

2.5 Instrumentation, Columns, and Software

1. Microplate reader (Epoch).
2. Gilson GX-274 ASPEC™ 4-probe positive pressure automated SPE system with 406 Dual Syringe Pumps (Gilson, Middleton, WI).
3. Proteomics off-line fractionation: Agilent 1200 HPLC System equipped with a quaternary pump, degasser, diode array detector, peltier-cooled autosampler, and fraction collector (set at 4 °C) (Agilent, Santa Clara, CA).
4. Proteomics off-line fractionation column: XBridge C-18 reversed-phase (RP) HPLC column, 250 mm × 4.6 mm, containing 5-μm particles, and a 4.6 mm × 20 mm guard column (Waters, Milford, MA).
5. Proteomics LC-MS: M-Class nanoAcquity dual pumping UPLC (Waters, Milford, MA) coupled with a Q Exactive HF Hybrid Quadrupole-Orbitrap MS (Thermo Scientific, San Jose, CA).
6. Proteomics LC column: Jupiter 3 μm and 5 μm C-18 media (Phenomenex, Torrance, CA), a 70-cm length of 360 μm o.d. × 75 μm i.d. and 4-cm length of 360 μm o.d. × 150 μm i.d. of fused silica (Molex, Lisle, IL).
7. Proteomics MS additional part 1: Kasil-based frits (Next Advance, Inc., Troy, NY).
8. Proteomics MS additional part 2: Nanospray Flex Ion Source (Thermo Scientific, San Jose, CA).
9. Metabolomics Gas Chromatography MS (GC-MS): Agilent 7890A GC coupled with a single quadrupole 5975C MS (Agilent Technologies, Inc., Santa Clara, CA).
10. Metabolomics GC column: HP-5MS or DB-5MS GC capillary column 30 m × 0.25 mm × 0.25 μm (Agilent, Santa Clara, CA).
11. Lipidomics LC-MS: H-Class nanoAcquity dual pumping UPLC (Waters, Milford, MA) coupled with a Velos Pro Orbitrap MS (Thermo Scientific, San Jose, CA).

12. Lipidomics LC column: Analytical LC column CSH 3.0 mm × 150 mm × 1.7 μm particle size (Waters, Milford, MA).
13. Proteomics software 1: MSCovert for spectrum to peak list conversion [28].
14. Proteomics software 2: mzRefinery for mass recalibration [29].
15. Proteomics software 3: MS-GF+ for peptide/protein identification [30].
16. Proteomics software 4: MASIC for extracting intensity values of TMT reporter ions [31].
17. Metabolomics software: MetaboliteDetector: comprehensive analysis tool for targeted and nontargeted GC-MS-based metabolome analysis [32].
18. Lipidomics software 1: LIQUID: an open-source software for identifying lipids [33].
19. Lipidomics software 2: MZmine: framework for processing, visualizing, and analyzing mass spectrometry-based molecular profile data [34].

3 Methods

MERS-CoV-infected cells are first lysed and extracted into metabolites, proteins, and lipids using MPLEx under BSL3 containment. The three analytes (in separate phases) are then individually collected into different tubes and vials, and this is the point of pathogen inactivation. The separate tubes are then decontaminated and brought out of the biological safety cabinet.

The protein phase is tryptically digested, desalted with C-18 columns, TMT labeled (and combined), separated via off-line C-18 RPLC (reversed-phase LC), concatenated, and analyzed by an M-Class nanoAcquity dual pumping UPLC coupled with a Q Exactive HF Hybrid Quadrupole-Orbitrap MS. The subsequent datasets are analyzed using MSCovert for spectrum to peak list conversion, mzRefinery for mass recalibration, and MS-GF+ for peptide/protein identification [28–30]. Quantitative information is extracted from the TMT reporter ion intensities using MASIC [31].

The metabolite phase is dried in vacuo, derivatized, and analyzed via GC-MS, and detected peaks are matched against an appropriate library to make metabolite identifications.

The lipid phase is dried in vacuo, reconstituted in methanol, and analyzed via a Waters Acquity UPLC H-class system and Velos Pro Orbitrap mass spectrometer, and lipid identifications are made via the LIQUID software [33]. MZmine is then used to process, visualize, and analyze peaks corresponding to identified lipids [34].

3.1 MPLEx Sample Extraction Under BSL3 Conditions

All work at BSL3 must be performed within a certified biological safety cabinet in a room under negative pressure with dedicated supply and exhaust using all appropriate personal protective equipment (PPE).

1. BSL3 Lysis: For each well in a 6-well plate, remove the media (disposing into a tray filled half way with 2× CiDecon™ phenolic disinfectant solution) and rinse cells with 3 mL of ~−40 °C Rapid Quenching Solution. Remove as quickly as possible and discard as above.
2. Add 150 μL of ice-cold 150 mM ammonium bicarbonate buffer to each well and scrape cells from plate. Rinse well with buffer to make sure all cells are collected. Remove to fresh 1.7 mL Sorenson tube (*see* **Notes 8** and **9**).
3. Add 600 μL −80 °C 2:1 chloroform/methanol (v/v) solution (fourfold excess) and close the tube tightly.
4. Shake sample vigorously for 10 s.
5. Incubate on ice for 5 min.
6. Shake sample vigorously for 10 s (*see* **Note 10**). This is the stage of pathogen inactivation.
7. Centrifuge at 13,000 × *g* for 10 min (in a centrifuge with a biocontainment lid) to separate the three phases (*see* Fig. 1).
8. Remove the upper phase to a fresh 1.7 mL Sorenson tube that has been labeled with the same sample name and "metabolites."
9. With a fresh tip, go through the protein "disc" at the interface, lightly push out any protein that may have entered the tip and bring the tip to the bottom of the tube to collect the organic phase. Stop collecting well before you reach the interface. Place the sample into a conical glass vial that has been labeled with the same sample name and "lipids."
10. Freeze aqueous and organic phases to bring out of BSL3 containment.

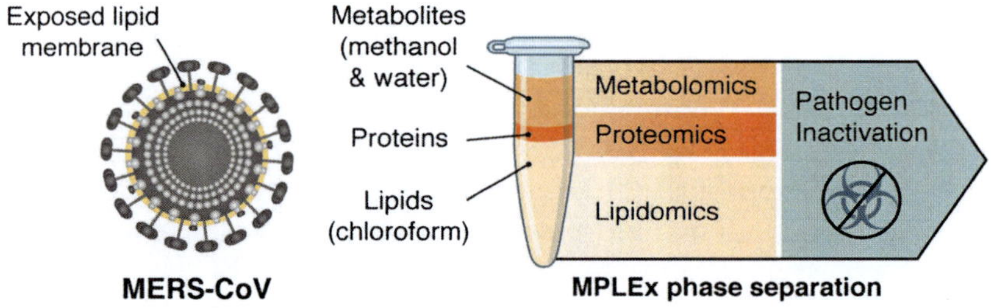

Fig. 1 Schematic of the MPLEx phase separation and pathogen inactivation. Repurposed from Ref. [6] with permission from The Royal Society of Chemistry

11. Add 200 μL 100% methanol to the protein "disc" and shake vigorously to mix. Pellet gently at 9000 × *g* for 5 min.
12. Gently decant the supernatant into waste and allow the protein pellet to "semi" dry upside down standing on a Kimwipe in the biological safety cabinet for 5 min.
13. Freeze at −80 °C prior to removing from BSL3.
14. Surface decontaminate all samples with 70% ethanol, seal into secondary containment and remove from facility. Return to −80 °C freezer at BSL2.
15. Dry in vacuo all aqueous phase and organic phase samples. Open tube and vial lids and place samples into slots. Use a no-heat setting and check progress after 2.5 h. Remove desiccated samples and replace with remaining ones until all samples have been dried and return desiccated samples to −80 °C prior to either shipping or further analysis (*see* **Note 11**).
16. The lightly dried protein pellet is stored at −80 °C until processing (*see* Subheadings 3.2, 3.5, and 3.7).
17. The dry metabolite samples are stored at −20 °C until processing (*see* Subheadings 3.3, 3.4, and 3.8).
18. The dry lipid samples are reconstituted in 500 μL of 2:1 chloroform:methanol and stored at −20 °C with lids that do not have a pre-slit septa (to reduce evaporation) until processing (*see* Subheadings 3.6 and 3.9).

3.2 Protein Tryptic Digestion, TMT Isobaric Labeling and High-Performance Liquid Chromatography (HPLC) Fractionation

1. Denature and Reduce Protein:
   (a) Add up to 200 μL of 8 M urea solution to the protein pellet for protein denaturation.
   (b) Using the BCA reagent kit, perform a protein assay according to the manufacturer's instructions. The protein assay is used to determine the initial protein mass of the sample prior to digestion (*see* **Note 12**).
   (c) Add enough 0.5 M DTT to get a final concentration of 5 mM for protein reduction.
   (d) Vortex and lightly water-bath sonicate the protein into solution.
   (e) Incubate at 60 °C for 30 min on a Thermomixer with a ThermoTop at 1000 rpm constant shaking.
2. Alkylate:
   (a) Add enough 0.4 M IAM to each sample to get a final concentration of 40 mM.
   (b) Incubate for 1 h at room temperature, protected from light, on a Thermomixer with a ThermoTop at 1000 rpm constant shaking.

3. Digest:
   (a) Dilute each sample tenfold with NH_4HCO_3, pH 8.4.
   (b) Add enough 1 M $CaCl_2$ to reach a final concentration of 1 mM.
   (c) Incubate 1 μg/μL trypsin for 15 min at 37 °C to activate.
   (d) Add trypsin to each sample at a 1:50 (w/w) enzyme-to-protein ratio and incubate at 37 °C for 3 h on a Thermomixer with a ThermoTop at 700 rpm constant shaking.
   (e) Flash freeze with liquid nitrogen and store at −80 °C until solid phase extraction (SPE) desalting can be performed.
4. C-18 SPE Desalting:
   (a) Thaw the sample and centrifuge at 10,000 × *g* for ~10 min to remove any precipitate and transfer to a fresh tube making sure not to pull up anything that might have pelleted.
   (b) Use one 50 mg/1 mL Supelco C-18 column per sample (capacity = 2.5 mg). C-18 SPE desalting can be performed manually on a vacuum manifold or using an automated SPE system.
      - Condition: 3× 1 mL methanol.
      - Equilibrate: 3× 1 mL 0.1% TFA.
      - Pass sample over column.
      - Wash: 4× 1 mL 95:5H_2O:ACN, 0.1% TFA.
      - Allow column to go to dryness.
      - Elute: 1× 1 mL 80:20 ACN:H_2O, 0.1% TFA into clean 1.5-mL microcentrifuge tube.
   (c) Concentrate in vacuo to approximately 50 μL (1.5–2 h).
   (d) Centrifuge at 10,000 × *g* for 2 min, perform BCA assay, and determine volume.
   (e) Aliquot an equal mass of peptide (30–100 μg each) into separate tubes and dry down the peptides (*see* **Note 13**).
   (f) Add 30 μL of the TMT dissolution buffer to each sample and ensure the pH is between 7 and 9 using pH paper.
5. TMT Labeling:
   (a) Post-digestion TMT labeling is performed according to the manufacturer's instructions.
   (b) Bring each needed vial of TMT Reagent to room temperature and spin to collect the liquid at the bottom.
   (c) Add 41 μL of TMT resuspension reagent to each TMT label, vortex, and spin. Allow to dissolve for 5 min with occasional vortexing.

(d) Add each TMT label into each sample.

(e) Incubate for 1 h at room temperature.

(f) Add 8 μL of TMT reaction quenching buffer to the sample and incubate for 15 min to quench the reaction.

(g) Combine all of the samples per set together and dry down to ~100 μL to remove the ACN.

(h) Perform C-18 SPE clean up in order to remove excess tag from the sample as described previously (*see* Subheading 3.2, **step 4a–d**).

(i) Pipette the sample into an HPLC vial.

6. Reversed-Phase High-Performance Liquid Chromatography (RPLC):

   The TMT-labeled sample is separated on a Waters reversed-phase XBridge C-18 column (250 mm × 4.6 mm column containing 5-μm particles, and a 4.6 mm × 20 mm guard column) using an Agilent 1200 HPLC System.

   (a) Reconstitute the sample up to 930 μL with HPLC Solvent A and inject onto the column at a flow rate of 0.5 mL/min.

   (b) After sample loading, the C-18 column is washed for 35 min with solvent A, before applying a 90-min LC gradient with solvent B. The LC gradient starts with a linear increase of solvent B to 10% in 10 min, then a linear increase to 20% B in 15 min, and 30 min to 30% B, 15 min to 35% B, 10 min to 45% B, and another 10 min to 100% solvent B. The flow rate is 0.5 mL/min [35].

   (c) Without collecting the first 15 min, a total of 96 fractions are collected into a 96-well plate throughout the LC gradient in equal time intervals.

   (d) At this point the plate can be frozen or dried down in vacuo.

   (e) After the fractions are dried, each fraction is reconstituted in 100 μL of 50% MeOH. Concatenate the 96 fractions into 24 samples by combining every other row [35], and concentrate again in vacuo to remove MeOH in the samples.

   (f) Determine the peptide concentration of each fraction using the BCA protein assay kit and dilute each of them to 0.1 μg/μL with H_2O. Each fraction is analyzed using LC–MS/MS (*see* Subheadings 3.5 and 3.7).

3.3 Derivatization of Metabolites

1. Make a methoxyamine solution with a pyridine concentration of 30 mg/mL.
2. Dry aqueous metabolites briefly (30 min) in vacuo after removing from the freezer.

3. Perform methoximation by adding 20 μL of methoxyamine solution to the sample vial and vortex for 30 s at setting 5 on a vortexer. Use a bath sonicator to ensure the sample is completely dissolved.
4. Incubate the sample in a Thermomixer maintained at 37 °C for 1 h and 30 min with 1000 rpm shaking.
5. Invert the vial one time to mix the samples with condensed drops at the cap surface.
6. Spin the sample down for 1 min at 1000 × *g* at room temperature in a centrifuge.
7. Perform silylation by adding 80 μL of *N*-methyl-*N*-trimethylsilyltrifluoroacetamide and 1% trimethylchlorosilane (MSTFA) solution using a syringe and vortex for 10 s.
8. Incubate the sample in a Thermomixer maintained at 37 °C for 30 min with 1000 rpm shaking.
9. Invert the vial one time to mix the samples with condensed drops at the cap surface.
10. Spin the sample down for 5 min at 2000 × *g* at room temperature in a centrifuge.
11. Transfer the reacted solution into a small volume insert and return to the same vial.
12. Prepare a process blank sample using only reagents and FAMEs [27] as retention time alignment compounds (alternatively, a mixture of n-alkanes can be used) using the same protocol for the sample preparation.
13. Tighten the caps and add to the autosampler tray for GC-MS analysis (*see* Subheading 3.4).

3.4 Metabolomics GC-MS Instrumental Analysis

1. Install a nonpolar GC column in the GC oven (HP-5MS or DB-5MS, or similar polarity −30 m × 0.25 mm × 0.25 μm).
2. Tune and calibrate the mass spectrometer (MS) before analysis to make sure the machine records the MS data correctly and check the helium gas pressure for possible leakage.
3. Set up the optimized running parameters for the GC; injector: temperature at 250 °C, pressure and flowrate: 1 mL/min (or user defined value); oven temperature, ramping rate, holding time and the temperature of the MS transfer line should all be set at what was used to analyze the standard compounds used for constructing the reference database (if an in-house library is used).
4. Set up the optimized running parameters in the MS; mass scan range (50–600 *m/z*), ion source temperature (250 °C), and ionization energy (70 eV) (the parameters which were used to run the standard compounds to build the database).

5. Clean the injection syringe and plunger with acetone to remove any residue.
6. Transfer the derivatized samples to the sample tray and place them in a randomized order to minimize instrumental artifacts.
7. For large sample sets requiring more than 24 h to be analyzed, batching should be performed. Analyzing batches requires additional randomization and QC samples and/or internal standards to account for any type of variation that could be introduced through time. Necessary blanks and retention time standards need to be included in each batch being run on subsequent days.

3.5 Proteomics LC-MS/MS Instrumental Analysis

The LC-MS conditions presented in this section represent a typical method for analysis of isobaric labeled peptides (TMT-10plex™ Isobaric Labeling Reagent, Thermo Scientific). Use of this specific labeling reagent requires a MS capable of achieving >30k mass resolution in order to baseline resolve all reporter ions. However, other isobaric labeling reagents are available that do not require this level of resolution for labs not possessing instrumentation with this capability. Isobaric labeling strategies provide several benefits including multiplexing of large cohorts to improve throughput; use of a reference channel for data normalization; and improved quantitation due to elimination of run-to-run variation and reduction in sample processing variability. However, the data also suffer from “compression,” where signals from co-eluting peptides can affect reporter ion accuracy. There are several ways to mitigate this compression effect and one of the primary means is off-line fractionation of samples to reduce sample complexity. Another method uses the additional specificity of MS^3 analysis (e.g., Thermo Scientific’s SPS MS3 workflow) where interference is reduced through a second MS/MS of primary fragment ions from the first MS/MS scan. This MS^3 technique does not generally preclude the need for off-line fractionation. Interference can still occur, although it is generally minimized because the product ions selected for MS^3 are more likely to be unique to a single peptide. The disadvantage of using MS^3 is a significant hit on instrument duty cycle. It simply requires more time to perform the additional MS/MS event on primary fragment ions from every precursor ion. The method presented here uses only MS/MS, favoring an increased peptide identification rate over reduction in compression. For the purpose of this section, parameters for an M-Class nanoAcquity dual pumping UPLC (Waters, Milford, MA) coupled with a Q Exactive HF Hybrid Quadrupole-Orbitrap MS are described.

1. UPLC Setup:

   Prepare analytical and trapping columns in-house by slurry packing Jupiter 3 μm and 5 μm C-18 media into a 70-cm length of 360 μm o.d. × 75 μm i.d., and 4-cm length of 360 μm o.d. × 150 μm i.d. of fused silica, respectively. Use Kasil-based frits for retention of packing material.

(a) Install the analytical and trap columns in a configuration such that once the sample has been trapped and washed, it is eluted in the opposite direction onto the analytical column.

(b) Prepare mobile phases (MP); MP A: 0.1% formic acid in water and MP B: 0.1% formic acid in acetonitrile.

(c) Set injection volume to 5 μL using a 5 μL sample loop and a 1.2× overfill value, resulting in a total of 7 μL of sample being pulled from the autosampler vial for each injection.

(d) Set the trapping flow rate at 5 μL/min and a total trap time of 8 min.

(e) Set the analytical flow rate to 300 nL/min with the following gradient and wash profile (min:%B); 0:1, 8:1, 10:8, 28:12, 83:30, 105:45, 108:95, 118:95, 122:50, 124:95, 126:1, 128:50, 130:50, 132:1, 152:1.

(f) Events (min:event); 8:end trapping and start gradient, 28: start data acquisition, 108:end gradient and start column wash, 132:end column wash, 148:end data acquisition, 152:end column re-equilibration.

2. Coupling UPLC to MS:

   Couple the UPLC with the MS using a Nanospray Flex Ion Source, attaching the standard metal emitter tip that comes with the source to the end of the analytical column.

3. MS Setup:

   (a) Perform required cleaning and/or calibrations. A full calibration is completed after the instrument has been deep cleaned (all optics cleaned up to and including the selection quadrupole). Otherwise, only mass calibrations are performed as needed (discussed in the QC section to follow).

   (b) Create a tune file that uses an ion transfer tube temperature of 325 °C and a spray voltage of 2 kV.

   (c) Create a data-dependent exclusion method with precursor settings using; mass range *m/z* 300–1800, resolution 60k, AGC target 3e6, max ion time 20 ms, and spectrum data type profile. For MS/MS settings, loop count 12, isolation width 0.7, resolution 30k, AGC target 1e5, max ion time 100 ms, fixed first mass *m/z* 110, normalized collision energy 30, and spectrum data type centroid. Additional settings include: min AGC target 5×10^3; intensity threshold 5×10^4; charge exclusion unassigned and 1; peptide match preferred; exclude isotopes on; dynamic exclusion 30 s; if idle do not pick others. Total acquisition time 120 min. Assume any other parameters listed were left with default settings.

4. Quality Control (QC):
   (a) Add QC standard runs in a matter consistent with the goals of your research and the expected quality of your samples. If, during the course of method development, the samples are relatively "dirty," then consider running QC standards daily or where reasonable between batches or blocks.
   (b) Select a QC standard that is similar to the samples that will be analyzed and assure enough material is on hand to conduct the entire study. For example, a 125 ng injection of RAW 264.7 macrophage cells (whole cell tryptic digest) would serve as a good QC standard for most mammalian cell studies (*see* **Note 14**).
   (c) Once a QC baseline has been established, run sample batches and blocks such that they are bracketed by QC standard runs. Only sample runs bracketed by acceptable QC standard runs are to be used (*see* **Note 15**).
   (d) Visually inspect each run, looking for shifts in retention time of common peaks and changes in signal stability and intensity.
   (e) Monitor mass accuracy by calculating the mass error for a commonly observed contaminant (e.g., polysiloxane) that produces a singly charged ion having a mass (m/z) of 445.1200.
   (f) QC samples should be run at least weekly, between sample batches, or anytime instrument performance is in question.

3.6 Lipidomics LC-MS/MS Instrumental Analysis

1. The specific conditions listed below are for a Waters Acquity UPLC H class system and Velos Pro Orbitrap mass spectrometer (*see* **Note 16**).
2. Set up and purge the LC system with lipid mobile phase A and B for 2 min or until all lines within the flow path have been cleared.
3. Install an analytical LC column (Waters CSH 3.0 mm × 150 mm, 1.7 μm particle size) and set the column temperature at 42 °C and the injection volume to 10 μL.
4. Set the flow rate to 250 μL/min and build the gradient profile as follows: (min, %B): 0, 40; 2, 50; 3, 60; 12, 70; 15, 75; 17, 78; 19, 85; 22, 92 25, 99; 34, 99; 34.5, 40.
5. Tune and calibrate the mass spectrometer according to the manufacturer's recommendations.

6. Set up the source parameters as follows: MS inlet and source 350 °C, spray voltage of 3.5 kV, sheath gas flow of 45, auxiliary gas flow of 30, and sweeping gas flow of 2.
7. Analyze each sample in both positive and negative ion mode with a precursor scan range of 200–2000 *m/z* and at a mass resolution of 60k.
8. Set up a data-dependent MS/MS method consisting of fragmentation of the top four ions, alternating between HCD and CID modes, and with the following dynamic exclusion parameters: repeat count 1, repeat duration 15 (s), exclusion list size of 250, exclusion duration of 8 (s), a low exclusion mass width of 0.55 and a high exclusion mass width of 1.5.
9. Acquire CID spectra using a minimum signal threshold of 500, isolation width of 2 *m/z*, normalized collision energy of 35, default charge state of 2 (*see* **Note 17**), activation Q value of 0.18 and activation time of 10.0 (ms).
10. Acquire HCD spectra at a resolution of 7500, a minimum signal threshold of 1000, isolation width of 2 *m/z*, normalized collision energy of 30, default charge state of 2, activation time of 0.100 (ms), and the first mass fixed at 90 *m/z*.
11. Couple the LC to the mass spectrometer and acquire three QC samples as a means of checking LC and instrument performance. The BTLE QC sample is injected and analyzed together with the samples as a means to monitor system performance, such as retention time, peak intensity, peak shape, and mass measurement error using data acquisition software. Furthermore, the BTLE QC sample is injected and analyzed after cleaning and calibration of the instrument, replacing the analytical column or any maintenance performed on the system.
12. Add the following sequence to the beginning and end of the sample queue to monitor carryover and instrument performance: methanol blank, QC, methanol blank.
13. For large sample sets insert the sequence from **step 12** throughout the queue as necessary.
14. Total Lipid Extracts (TLEs) are stored in 2:1 chloroform: methanol and evaporated in vacuo, then reconstituted in 5 μL chloroform and 45 μL of methanol prior to injecting 10 μL onto the column.

3.7 Protein Identifications

For the proteomics data analysis, there is a variety of commercial and open-source tools currently available. The analysis parameters are also by the preference of the user. In our laboratory, we regularly perform proteomic analysis with a combination of msConvert (for spectrum to peak list conversion) [28], mzRefinery (for mass recalibration) [29], MS-GF+ (for peptide/protein identification) [30], and MASIC (for quantification) [31].

1. Peak list conversion with msConvert parameters:
   (a) Convert to _DTA.txt format.
   (b) Convert to centroid peak.
2. Mass recalibration with mzRefinery parameters:
   (a) Same searching parameters described below.
3. Peptide/protein identification with MS-GF+:
   (a) Protein sequence database. The database can be downloaded from a variety of sources, such as RefSeq, UniprotKB, and Ensembl. For human cells, we recommend using Swiss-Prot, which is available at www.uniprot.org.
   (b) Searching parameters:
      - Enzyme: Trypsin.
        - Number of tolerable termini: 1 (partial trypsin digestion).
        - Minimum peptide length: 6.
        - Maximum Peptide Length: 50.
        - Number of matches per spectrum: 1.
      - Variable modifications: methionine oxidation (+15.9949 Da).
      - Fixed modifications: cysteine carbamidomethylation (+57.0215).
      - Parent mass tolerance: 20 ppm.
      - Isotope error: −1 and 1.
      - Targeted Decoy Analysis: true.
      - Fragmentation method: automatic detected (as written in the spectrum).
      - Instrument: QExactive.
      - Filter searching results based on MS-GF probability score. Usually, a score $\leq 1 \times 10^{-10}$ results in a false-discovery rate $\leq 1\%$.
      - Extraction of quantitative information with MASIC. Parameters:
        - Mass tolerance for reporter ions: 0.003 m/z.
        - The identified peptides and reporter ion intensities can be combined using Microsoft Access. Sum the intensities of multiple spectra of the same peptide and multiple peptides of the same protein.

- Expected results: the number of identified peptides and proteins will vary according to the mass spectrometer and peptide separation strategy. For instance, mammalian cells infected with MERS-CoV prefractionated into 24 fractions and analyzed in a 100-min gradient on a QExactive mass spectrometer should lead to the identification of 8000–10,000 proteins.
- Troubleshooting: If the number of identified proteins is much smaller than expected, check the following items:
  - Contamination: detergents and plasticizers are common contaminants in samples and they have a characteristic profile in LC-MS/MS data by having uniform mass and retention time shifts. The solution to this problem is to eliminate the source of these contaminants by washing the glassware, avoiding detergents during sample preparation steps and checking the compatibility of used plastic tubes with the extraction solvents.
  - Searching parameters: another source of poor coverage can be due to incorrect searching parameters. Check the sequence database (if correct species was used, for instance), enzyme specificity, and peptide modifications.

3.8 Metabolite Identifications

1. Check that all of the data files were correctly obtained from the analysis. If an internal standard(s) was spiked, make sure the retention time and peak intensity values are consistent throughout the sample analysis.
2. Convert the MS raw data into a general MS format if required (e.g., netCDF).
3. Upload all the data files into the chosen software (e.g., Metabolite Detector, MassHunter, AMDIS, or similar) and perform an alignment of retention time using the acquired data from the analysis of retention time alignment compounds (e.g., FAMEs or Alkane mix).
4. Align the retention time information to all the collected data files, and perform library matching in order to identify metabolites.
5. Perform a batch analysis to align all the detected metabolite peaks.
6. Export the processed data into a CSV format, and read the data from a spreadsheet program (e.g., Excel).
7. Check the matching scores of identified metabolites and their signal-to-noise ratios (higher than 75/100 matching score, with $S/N > 3$). Select the correctly identified metabolites for

further biological interpretation (e.g., integrated analysis with other omics data including proteomics and lipidomics).

3.9 Lipid Identifications

1. Launch LIQUID program [33] and upload a .raw or .mzXL data file using the drop down menu.
2. Select the ionization mode the samples were acquired in and enter mass error tolerances based on instrument QCs and sample data.
3. Load lipid targets, set results per scan to 1, then process targets.
4. Sort the list of potential lipid identifications by common name, and then scan number.
5. Inspect the potential identifications one at a time, examining and evaluating the following:
   (a) Diagnostic ion and associated fragment ions (e.g., fatty acids). Verify that the expected peaks are present.
   (b) Isotopic profile: it should match predicted profile.
   (c) Extracted ion chromatogram: confirm the parent scan is at the peak apex.
   (d) Mass error: confirm it is consistent with the rest of the data.
   (e) Retention time: verify that it is within allowable tolerances for LC system.
6. Select the correct identifications as going through **step 5.**
7. Export selected results.
8. Using LIQIUD's exported results from **step** 7, create a target list including observed *m/z*, retention time, and common name.
9. Launch MZmine and import data files.
10. Associate targeted files from **step 8** to the raw data files imported in **step 9**.
11. Perform mass detection on files not analyzed in LIQUID, and then build chromatograms from generated mass list.
12. Perform peak alignment and then gap-filling on the aligned list.
13. Check feature alignment and manually verify correct alignment for each lipid identification.
14. Export final results including lipid name and peak area or peak apex intensity, which can be used for statistical analysis.

4 Notes

1. 18.2 MΩ cm water (which will be referred to as water) should be used to prepare all solutions unless otherwise noted.

2. The chloroform:methanol solution should always be at least −20 °C to facilitate protein precipitation.
3. Order the highest purity possible.
4. Purchase MSTFA with 1% TMCS in a 5 mL size. Once it is opened, moisture in the air can reduce the reactivity of the reagent. It can cause skin corrosion, serious eye damage, and specific target organ toxicity, and is a flammable liquid and vapor. Wear safety glasses, gloves, and lab coat, and work in a fume hood.
5. All tryptic digestion chemicals are purchased from Sigma-Aldrich unless otherwise listed.
6. Lysine residues modified by reductive methylation, yielding a highly active and stable trypsin.
7. Polypropylene is conditionally resistant to chloroform and can be used as long as contact time is minimized. A fresh tip must be used each time after chloroform is dispensed.
8. The Sorenson brand tubes have shown to be compatible with chloroform and do not leach polymers that can interfere with downstream analysis.
9. Three sets of Sorenson tubes should be generated for each sample. Samples can be collected and MPLExed in the tubes labeled for protein/proteomics as the protein remains in the tube following removal of the aqueous (into a tube labeled metabolites) and organic (into a tube labeled lipids) phases (*see* Fig. 1).
10. Aliquots of samples are removed and tested at this step to ensure all virus has been inactivated.
11. At this point the sample is inactivated and separated into three distinct analytes and each of them can be handled in BSL1 conditions.
12. BCA assay regent is compatible up to 3 M urea, so when the sample is in 8 M urea, appropriate dilution factors should be considered.
13. The TMT Reagents are amine-reactive and modify lysine residues and the peptide N-termini. All amine-containing buffers and additives must be removed before digestion and labeling.
14. Prior to beginning a study, determine acceptable values for a selected set of metrics for the QC standard to be used. This should be performed using a clean instrument and new analytical and trap columns. Example metrics to monitor for performance are number of unique peptides identified, average chromatographic peak width, and mass error. QC procedures for complex proteomics measurements are not always

straightforward. Establish a baseline for the QC process before running samples.

15. Always run a blank before and after QC standard runs.
16. In general, an LC capable of delivering a flow rate of 250 μL/min with a pressure limit of 8000 psi or greater and a mass spectrometer with a mass resolution of 30k or higher is required for this method.
17. The default charge state of 2 is used instead of 1 to ensure we don't miss the doubly charged lipids (e.g., cardiolipins and gangliosides) and the detection of single charged lipids by setting the default charge state to 2 is unaffected.

Acknowledgments

The methods described in this chapter were supported in part by National Institute of Allergy and Infectious Diseases grant U19AI106772. Additional support was provided by National Institute of General Medical Sciences Grant P41 GM103493. Lipidomics, proteomics, and metabolomics analyses were performed in the Environmental Molecular Sciences Laboratory, a national scientific user facility sponsored by the Department of Energy (DOE) Office of Biological and Environmental Research, and located at Pacific Northwest National Laboratory (PNNL). PNNL is a multiprogram national laboratory operated by Battelle for the DOE under Contract DE-AC05-76RLO 1830.

References

1. Modjarrad K et al (2016) A roadmap for MERS-CoV research and product development: report from a World Health Organization consultation. Nat Med 22(7):701–705
2. Aderem A et al (2011) A systems biology approach to infectious disease research: innovating the pathogen-host research paradigm. MBio 2(1):e00325–e00310
3. Cockrell AS et al (2016) A mouse model for MERS coronavirus-induced acute respiratory distress syndrome. Nat Microbiol 2:16226
4. Menachery VD et al (2018) MERS-CoV and H5N1 influenza virus antagonize antigen presentation by altering the epigenetic landscape. Proc Natl Acad Sci U S A 115(5): E1012–E1021
5. Nakayasu ES et al (2016) MPLEx: a Robust and universal protocol for single-sample integrative proteomic, metabolomic, and lipidomic analyses. mSystems 1(3):e00043
6. Burnum-Johnson KE et al (2017) MPLEx: a method for simultaneous pathogen inactivation and extraction of samples for multi-omics profiling. Analyst 142(3):442–448
7. Menachery VD et al (2014) Pathogenic influenza viruses and coronaviruses utilize similar and contrasting approaches to control interferon-stimulated gene responses. MBio 5 (3):e01174–e01114
8. Mitchell HD et al (2013) A network integration approach to predict conserved regulators related to pathogenicity of influenza and SARS-CoV respiratory viruses. PLoS One 8 (7):e69374
9. Gralinski LE et al (2013) Mechanisms of severe acute respiratory syndrome coronavirus-induced acute lung injury. MBio 4(4):e00271
10. Sims AC et al (2013) Release of severe acute respiratory syndrome coronavirus nuclear import block enhances host transcription in human lung cells. J Virol 87(7):3885–3902

11. Kyle JE et al (2019) Plasma lipidome reveals critical illness and recovery from human Ebola virus disease. Proc Natl Acad Sci U S A 116 (9):3919–3928
12. Eisfeld AJ et al (2017) Multi-platform omics analysis of human ebola virus disease pathogenesis. Cell Host Microbe 22(6):817–829 e8
13. Tisoncik-Go J et al (2016) Integrated omics analysis of pathogenic host responses during pandemic H1N1 influenza virus infection: the crucial role of lipid metabolism. Cell Host Microbe 19(2):254–266
14. Nicod C, Banaei-Esfahani A, Collins BC (2017) Elucidation of host-pathogen protein-protein interactions to uncover mechanisms of host cell rewiring. Curr Opin Microbiol 39:7–15
15. Wang YP, Lei QY (2018) Metabolite sensing and signaling in cell metabolism. Signal Transduct Target Ther 3:30
16. DeBerardinis RJ, Thompson CB (2012) Cellular metabolism and disease: what do metabolic outliers teach us? Cell 148(6):1132–1144
17. Sanchez EL, Lagunoff M (2015) Viral activation of cellular metabolism. Virology 479:609–618
18. Wenk MR (2006) Lipidomics of host-pathogen interactions. Chem Phys Lipids 143 (1-2):42–42
19. Knoops K et al (2008) SARS-coronavirus replication is supported by a reticulovesicular network of modified endoplasmic reticulum. PLoS Biol 6(9):1957–1974
20. Knoops K et al (2010) Integrity of the early secretory pathway promotes, but is not required for, severe acute respiratory syndrome coronavirus RNA synthesis and virus-induced remodeling of endoplasmic reticulum membranes. J Virol 84(2):833–846
21. Ulasli M et al (2010) Qualitative and quantitative ultrastructural analysis of the membrane rearrangements induced by coronavirus. Cell Microbiol 12(6):844–861
22. de Wilde AH et al (2013) MERS-coronavirus replication induces severe in vitro cytopathology and is strongly inhibited by cyclosporin A or interferon-alpha treatment. J Gen Virol 94:1749–1760
23. Miller S, Krijnse-Locker J (2008) Modification of intracellular membrane structures for virus replication. Nat Rev Microbiol 6(5):363–374
24. Stapleford KA, Miller DJ (2010) Role of cellular lipids in positive-sense RNA virus replication complex assembly and function. Viruses 2 (5):1055–1068
25. Blanchard E, Roingeard P (2015) Virus-induced double-membrane vesicles. Cell Microbiol 17(1):45–50
26. Folch J, Lees M, Stanley GHS (1957) A simple method for the isolation and purification of total lipides from animal tissues. J Biol Chem 226(1):497–509
27. Kind T et al (2009) FiehnLib: mass spectral and retention index libraries for metabolomics based on quadrupole and time-of-flight gas chromatography/mass spectrometry. Anal Chem 81(24):10038–10048
28. Adusumilli R, Mallick P (2017) Data conversion with ProteoWizard msConvert. Methods Mol Biol 1550:339–368
29. Gibbons BC et al (2015) Correcting systematic bias and instrument measurement drift with mzRefinery. Bioinformatics 31 (23):3838–3840
30. Kim S, Pevzner PA (2014) MS-GF+ makes progress towards a universal database search tool for proteomics. Nat Commun 5:5277
31. Monroe ME et al (2008) MASIC: a software program for fast quantitation and flexible visualization of chromatographic profiles from detected LC-MS(/MS) features. Comput Biol Chem 32(3):215–217
32. Hiller K et al (2009) MetaboliteDetector: comprehensive analysis tool for targeted and non-targeted GC/MS based metabolome analysis. Anal Chem 81(9):3429–3439
33. Kyle JE et al (2017) LIQUID: an-open source software for identifying lipids in LC-MS/MS-based lipidomics data. Bioinformatics 33 (11):1744–1746
34. Pluskal T et al (2010) MZmine 2: modular framework for processing, visualizing, and analyzing mass spectrometry-based molecular profile data. BMC Bioinformatics 11:395
35. Wang Y et al (2011) Reversed-phase chromatography with multiple fraction concatenation strategy for proteome profiling of human MCF10A cells. Proteomics 11 (10):2019–2026

Chapter 15

Evaluation of Activation and Inflammatory Activity of Myeloid Cells During Pathogenic Human Coronavirus Infection

Rudragouda Channappanavar and Stanley Perlman

Abstract

Innate immune cells play a vital role in mounting an effective host response to a variety of pathogen challenges. Myeloid cells such as neutrophils and monocyte-macrophages are major innate leukocytes that orchestrate protective immunity to viral lung infections. However, a dysregulated cytokine response can promote excessive infiltration and robust pro-inflammatory activity of neutrophils and monocyte-macrophages, leading to fatal disease. Following virus infection, the beneficial or deleterious role of infiltrating neutrophils and monocyte-macrophages is determined largely by their ability to secrete inflammatory cytokines and chemokines. A majority of studies use the total number of infiltrating cells and their activation status as measures to demonstrate their role during an infection. Consequently, the ability of neutrophils and Inflammatory Monocyte Macrophages (IMMs) to secrete inflammatory cytokines and chemokines, and its correlation with the disease severity, is not well defined. In this chapter, we report useful markers to identify lung infiltrating innate immune cells and define their activation status. We also describe a simple method to measure intracellular cytokine production to evaluate the inflammatory activity of neutrophils and IMMs in a mouse model of human coronavirus infection.

Key words Coronavirus, Neutrophils, Inflammatory monocyte-macrophages, Lungs, Cytokines and chemokines

1 Introduction

Myeloid cells such as neutrophils and monocyte-macrophages are key immune cells that make up a large proportion of tissue infiltrating innate leukocytes following a pathogen challenge. Both neutrophils and inflammatory monocytes-macrophages (IMMs) are rapidly recruited to the site of infection and play crucial roles in the host defense against viral lung infections [1, 2]. The antiviral functions of neutrophils and monocyte-macrophages are facilitated following the recognition of pathogen-associated molecular patterns (PAMPs) by the cell surface and endosomal toll-like receptors (TLRs) and intracellular RIG-I like (RLRs) and Nod-like receptors

Rahul Vijay (ed.), *MERS Coronavirus: Methods and Protocols*, Methods in Molecular Biology, vol. 2099,
https://doi.org/10.1007/978-1-0716-0211-9_15,

(NLRs). Detection of viral PAMPs (viral proteins and nucleic acids) by these sensors leads to the activation of a cascade of signaling events resulting in the production of antiviral molecules like interferons (IFNs), interferon-stimulated genes (ISGs), and inflammatory cytokines and chemokines [3–5]. IMMs and neutrophils also participate in the phagocytosis of virus-infected cells and orchestrate effective adaptive T cell responses, both of which are essential for effective virus clearance [6].

In addition to host protective function of myeloid cells during viral lung infections, several recent studies demonstrate their role in mediating cytokine storm and thus exacerbating the host immune response to virus infections [2, 7]. The deleterious functions of neutrophils and IMMs are linked to dysregulated type I IFN (IFN-I) responses, particularly during high pathogenic virus infections [8, 9]. For example, while a controlled neutrophil response is protective during influenza A virus infection, an excessive neutrophil accumulation is detrimental [10, 11]. Similarly, an exaggerated monocyte-macrophage response resulting from delayed IFN-I signaling is detrimental during human coronavirus infections [8]. IMMs and neutrophils also express increased levels of death receptors such as DR5 and FAS, and the interaction of these receptors with their ligands TRAIL and FASL, respectively, promotes airway epithelial and lung microvascular endothelial cell death [9, 12, 13]. Additionally, excessive inflammatory cytokines and chemokines produced by IMMs and neutrophils impair antiviral T cell responses, leading to ineffective virus clearance and reduced survival [8].

A majority of the studies demonstrating the beneficial or detrimental effects of neutrophils and IMMs during viral lung infections enumerate percentages and total number and define activation status of the lung infiltrating myeloid cells using surface markers [14]. We recently showed spontaneous production of several inflammatory cytokines and chemokines by neutrophils and IMMs, which correlated with severe lung pathology and reduced survival in CoV infections [8]. Thus, the identification of specific inflammatory cytokines and chemokines produced by these cells will allow us to define their pro-inflammatory status and design strategies to control inflammatory responses. In this study, we describe useful markers to identify innate immune cells infiltrating into the lung and describe a simple method to evaluate inflammatory cytokine and chemokine production by neutrophils and monocyte-macrophages during pathogenic human coronavirus infections.

2 Materials

2.1 Infection

1. Sterile, endotoxin-free, pharmaceutical grade physiological saline, 1× phosphate-buffered saline, or 1× Dulbecco's Modified Minimum Essential Medium.
2. Isoflurane, isoflurane vaporizer and induction chamber or cocktail of xylazine (12.5 mg/kg) + ketamine (85 mg/kg).
3. 1 mL syringe for anesthetic administration.
4. Sterile 200 μL pipette tips and single channel 200 μL pipette.
5. Human coronavirus aliquots.

2.2 Harvesting Lungs

1. Anesthetic—ketamine (85 mg/kg) + xylazine (12.5 mg/kg).
2. 1 mL syringe for anesthetic administration and 10 mL syringe for lung perfusion.
3. Surgical equipment (scissors, forceps, 22G 5/7 needles).
4. 1× PBS or pharmaceutical grade physiological saline and 22G 5/7 needles
5. Square (6–12 inch) styrofoam or cardboard, absorbent pads.

2.3 Digestion and Processing of Lung Tissue

1. Scissors and forceps.
2. A 12-well plate for mincing lung tissue.
3. Lung digestion DNAse I/Collagenase D buffer.
4. 15 mL and 50 mL conical tubes.
5. Tube rotator.
6. Six-well plates for homogenizing lung tissue.
7. 3 mL syringe plunger, plastic Pasteur pipettes (3 mL).
8. RPMI 10% FBS media.
9. Benchtop lab centrifuge with rotors and cups to hold 15 mL or 50 mL conical.

2.4 Incubation and/or Stimulation of Lung Cells for Intracellular Cytokine Staining

1. 96-well plates, 200 μL multichannel pipette, 200 μL single channel pipette
2. RPMI 10% FBS, and Golgi-plug.
3. TLR ligands: Poly I:C, LPS, and R837.

2.5 FACS Staining and Acquisition

1. 96-Well plates.
2. FACS buffer (PBS+ 2–5% FBS+0.01% sodium azide).
3. Cytofix/Cytoperm buffer.
4. Perm/Wash Buffer.
5. Antibodies: anti-mouse CD45 PE-Cy7 (Clone: 30-F11), anti-mouse CD11b e450 Cat (Clone: M1/70), anti-mouse CD11c

PE (Clone: N418), anti-mouse IA/IE PerCp-Cy5.5 (clone: M5/114.15.2), anti-mouse Ly6C percp-cy5.5 (Clone: HK1.4), anti-mouse Ly6G FITC (Clone:1A8,), anti-mouse TNF APC (Clone: MP6-XT22), anti-mouse IL-6 APC (Clone:MP5-20F3), anti-mouse iNOS APC (Clone: CXNFT), anti-mouse IL-1β APC (Clone: NJTEN3), anti-mouse CD80 APC (Clone: 16-10A1), anti-mouse CD86 APC (Clone:GL-1\), anti-mouse CD69 APC (Clone: H1.2F3), anti-mouse PDCA-1 PE/APC (Clone:JF05-1C2.4.1), and anti-mouse CD16/32 (clone: 2.4G2).

6. Flow Cytometer (capable of detecting six or more fluorophores).

3 Methods

3.1 Mice Infection

1. Thaw a virus aliquot on ice just before infection, avoiding repeated freeze-thaw.
2. Dilute MERS-CoV and SARS-CoV in DMEM to achieve the required dose and keep virus on ice throughout the infection time (*see* **Note 1**).
3. Under xylazine/ketamine anesthesia (confirmed by pedal reflex), slowly deliver 40–50 μL of a well-mixed virus inoculum directly into the nostrils using a 200 μL pipette (*see* **Note 2**).
4. Following virus delivery, mouse should be placed on its dorsal side in a cage with bedding for the remaining virus inoculum to be inhaled.
5. Monitor mice every 10 min until complete recovery.

3.2 Harvesting and Digestion of Lungs

1. On days 4–5 post-infection, under complete xylazine/ketamine or isoflurane anesthesia (confirmed by pedal reflex), cut open the abdominal cavity to expose the diaphragm. Make an incision through the diaphragm with scissors, remove diaphragm, and then remove the rib cage to completely expose the heart and lungs (*see* **Note 3**).
2. Fill a 10 mL syringe with ice-cold sterile PBS and attach a 25 G × 5/8 needle. Insert the needle into the right ventricle of the heart and slowly inject 5 mL of DPBS into the heart. In the meantime, use forceps to break the left atria to allow blood to drain from circulation. Inject remaining 5 mL of DPBS until lungs turn pale.
3. Separate the heart and then remove the lung from the thoracic cavity. Separate and discard any remaining connective tissue associated with the lungs (*see* **Note 4**).
4. Place the lungs into the well of a 12-well tissue culture plate filled with 2.5 mL of DPBS on ice.

5. Rinse the lungs with DPBS and transfer it into another well without DPBS. Mince the lungs into very fine pieces using scissors.
6. Transfer minced lungs with a 2.5 mL plastic transfer pipette to a 15 mL conical tube containing 5 mL of digestion buffer.
7. Place tubes on a rocker and gently rock at room temperature for 30 min (*see* **Note 5**).
8. Place a 70 μm cell strainer in a 60 × 15 mm tissue culture dish or in a well of a 6-well plate.
9. Transfer lung tissue in digestion buffer on to the cell strainer using a 2.5 mL transfer pipette. Gently press and dissociate tissue through a strainer with the flat end of a 3 mL syringe plunger. Process tissues until there is only connective tissue remaining on the strainer and rinse the strainer with complete RPMI 1640 medium.
10. Spin down lung cells in 15 mL conical tube for 5 min at 300 × *g* at 4 °C in a bucket tabletop centrifuge.
11. Discard off supernatant and resuspend the cells in 1 mL of ACK buffer for 1 min to lyse the remaining red blood cells. Neutralize the ACK buffer with 10 mL of ice-cold DPBS or 10 mL of 5% RPMI medium.
12. Spin down the cells for 5 min at 300 × *g* at 4 °C and resuspend the cells in 5 mL of ice-cold buffer.

3.3 Cell Surface Staining for Innate Immune Cells (See Note 6)

1. Spin down lung cells for 5 min at 300 × *g* at 4 °C.
2. Dilute 0.2 μg of CD16/32 antibodies in 100 μL FACS buffer and resuspend cells in FACS buffer containing CD16/32 antibodies in a 96-well plate.
3. Gently mix the cells and antibodies.
4. Incubate the cells in the dark for 15 min at 4 °C.
5. Wash the cells twice with 150 μL of FACS buffer at 300 × *g* for 5 min at 4 °C.
6. Resuspend the cells in 100 μL FACS buffer containing the antibody cocktail.
7. Incubate the cells in the dark for 20 min at 4 °C.
8. Spin down lung cells for 5 min at 300 × *g* at 4 °C and discard the FACS buffer.
9. Wash the cells twice with 150 μL of FACS buffer at 300 × *g* for 5 min at 4 °C.
10. Resuspend the cells in 200 μL of FACS buffer and acquire using a flow cytometer.
11. Figure 1 provides an example of IMM activation marker expression in CoV infected lungs.

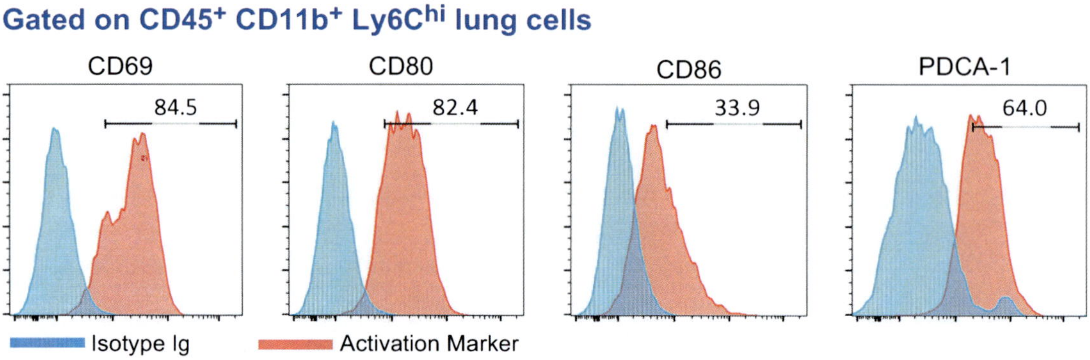

Fig. 1 Ideal markers to determine activation status of IMMs. Lung cells harvested from SARS-CoV-infected BALB/c mice (3 dpl) were surface stained for IMMs and activation markers as described in the methods section

3.4 Incubation and or Stimulation of Lung Cells for Intracellular Cytokine Staining (See Note 7)

1. Count live cells using a hemocytometer by staining with trypan blue.
2. Spin down lung cells for 5 min at 300 × *g* at 4 °C and resuspend the cells in RPMI 10% FBS at one million cells per 100 μL/well.
3. Dispense 100 μL of cells into a well of 96-well plates and add additional 100 μL RPMI 10% FBS media with or without Golgi-plug (1 μg/mL) with or without a TLR agonist (LPS 10–100 ng/mL, R837 and Poly I:C 100 ng to 1 μg/mL).
4. Incubate cells for 6–7 h at 37 °C in CO_2 incubator.
5. After incubation, wash cells twice with RPMI 10% FBS media.

3.5 Cell Surface and Intracellular Cytokine Staining for FACS (See Note 8)

1. Spin down lung cells for 5 min at 300 × *g* at 4 °C.
2. Dilute 0.2 μg of CD16 antibodies in 100 μL FACS buffer and resuspend cells in FACS buffer containing CD16/32 antibodies in a 96-well plate.
3. Gently mix the cells and antibodies.
4. Incubate the cells in the dark for 15 min at 4 °C.
5. Wash the cells twice with 150 μL of FACS buffer at 300 × *g* for 5 min at 4 °C.
6. Resuspend the cells in 100 μL FACS buffer containing 0.25 μg of cell surface identification and activation marker antibodies (*see* Table 1).
7. Incubate the cells in the dark for 20 min at 4 °C.
8. Spin down lung cells for 5 min at 300 × *g* at 4 °C and discard the FACS buffer.
9. Wash the cells twice with 150 μL of FACS buffer at 300 × *g* for 5 min at 4 °C.

Table 1
Cell surface markers to identify lung resident and lung infiltrating innate immune cells

No.	Innate immune cell	FACS markers
1	Alveolar macrophage	$CD45^+CD11c^+SiglecF^+$or $CD45^+CD11c^+F4/80^+$
2	Neutrophils	$CD45^+Ly6C^{int}Ly6G^+$or $CD45^+CD11b^+Gr1^+$
3	Monocytes	$CD45^+CD11b^+Ly6C^{hi}CCR2^+$
4	Macrophages	$CD45^+CD11b^+F4/80^+$
5	Dendritic cells	$CD45^+CD11b^-CD11c^+MHC\text{-}II^+$
6	Natural killer cells	$CD45^+CD3^-NKP46^+/CD45^+CD3^-NK1.1$ (B6)/$CD45^+CD3^-DX5^+$(BALB/c)
7	Eosinophils	$CD45^+CD11b^+CD11c^-SiglecF^+$

10. Add 100 μL of Cytofix/Cytoperm buffer and incubate cells in the dark for 25 min at 4 °C.
11. Add another 100 μL of 1× perm buffer (diluted to 1× with ddH_2O), spin down the cells (400 × *g* for 5 min at 4 °C), and discard the buffer.
12. Add 100 μL of 1× perm buffer containing anti-cytokine antibodies. See Table 1 for markers and concentrations.
13. Incubate the cells in the dark for 25–30 min at 4 °C.
14. After incubation add another 100 μL of 1× perm buffer (diluted to 1× with ddH_2O), spin down (400 × *g* for 5 min at 4 °C) the cells, and discard the buffer.
15. Wash the cells twice with 200 μL of 1× perm buffer at 400 × *g* for 5 min at 4 °C.
16. Wash the cells once with 200 μL of FACS buffer at 300 × *g* for 5 min at 4 °C (*see* **Note 9**).
17. Resuspend cells in 200 μL of FACS buffer for FACS acquisition.
18. Acquire FACS data using a flow cytometer and analyze data using FlowJo software (*see* **Note 10**).
19. Figure 2 demonstrates intracellular inflammatory cytokine production by lung IMMs on day 1 and 3 post-SARS-CoV infection.
20. Figure 3 demonstrates intracellular cytokine production by IMMs and neutrophils following brief TLR stimulation.

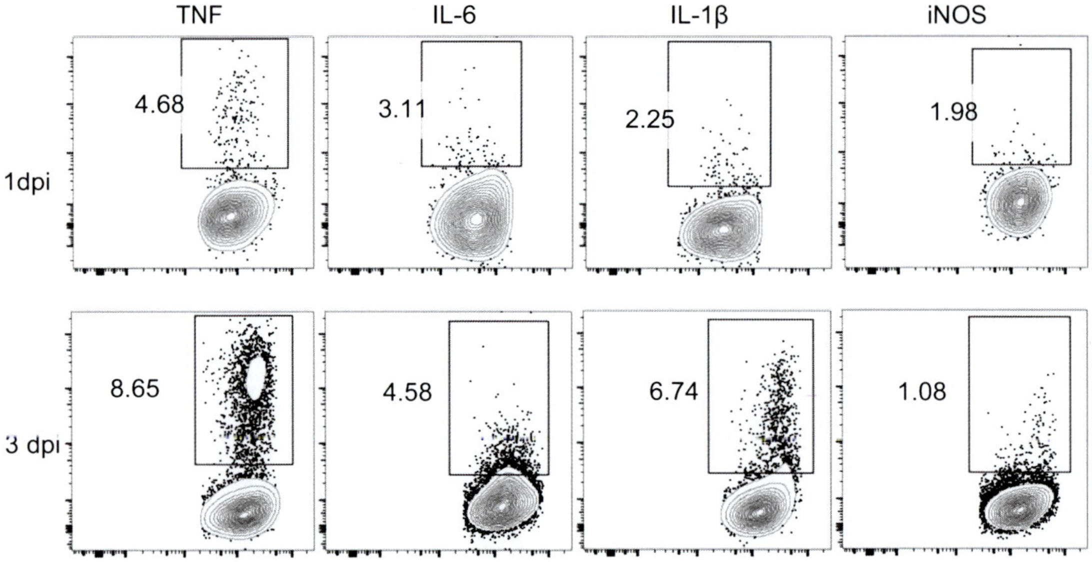

Fig. 2 Spontaneous cytokine production by IMMs. Lung cells isolated from SARS-CoV-infected BALB/c mice (1–3 dpi) were incubated for 7-h in the presence of Golgi-plug. Cells were then surface stained for IMMs ($CD45^{+}CD11b^{+}Ly6C^{hi}$) and then for intracellular cytokines TNF, IL-6, IL-1β, and iNOS

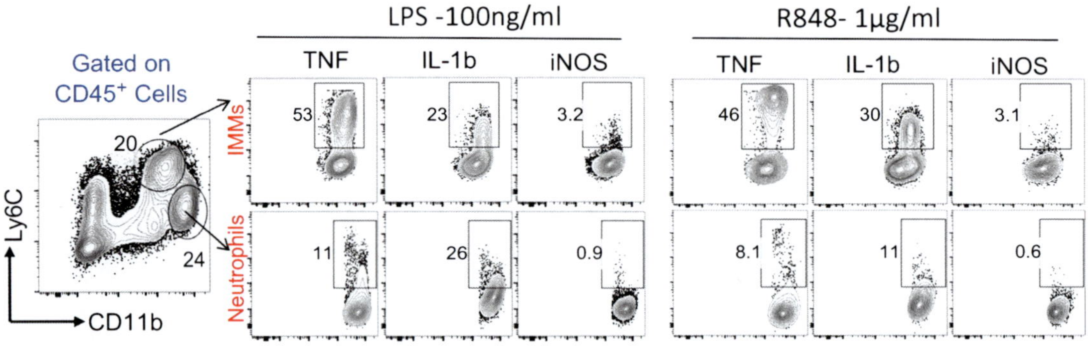

Fig. 3 Staining for intracellular cytokines in TLR-stimulated IMMs and neutrophils: Total lung cells isolated from SARS-CoV-infected BALB/c mice (3 dpi) were stimulated with LPS (TLR4 ligand, 100 ng/mL) or R848 (TLR7 ligand, 1 μg/mL) for 4—h in the presence of Golgi-plug. IMMs ($CD11b^{hi}Ly6c^{hi}$) and neutrophils ($CD11b^{hi}Ly6C^{int}$) were stained for intracellular TNF, IL-1β, and iNOS production

4 Notes

1. Both MERS-CoV and SARS-CoV should be aliquoted in ice-cold DMEM and the media should be kept on ice throughout the period of infection.
2. Ketamine/xylazine anesthesia gives uniform infection compared to isoflurane. It is essential to make sure that the animals

are completely anesthetized. Following anesthesia mice should be placed on their back on thick bedding to avoid hypothermia.

3. During MERS-CoV and SARS-CoV infection, neutrophil accumulation peaks between day 1 and 3 post-infection and monocytes between day 2 and 4 post-infection. As a result, days 2–4 are ideal time points to assess the activation and pro-inflammatory activity of myeloid cells.
4. Soon after harvesting, lungs should be placed in PBS on ice until further processed. Post-homogenization, the lung tissue can be digested at room temperature from 30 to 45 min.
5. After Collagenase/DNAse digestion, the lung tissue in 15 mL conical should be kept on ice until further processed. Following tissue homogenization, all washing and cell surface staining should be carried out on the ice or at 4 °C.
6. For efficient staining, antibodies should be diluted in 100 μL of FACS buffer and added to assigned well of 96-well round-bottom plate or tube. Both plate and tubes should be placed on flat surface of vortex machine (with very low speed) to ensure that cell pellet is broken and antibody solution is uniformly distributed.
7. The concentration of LPS for stimulation should be 10–100 ng/mL. Other TLR agonists could be used at 1 μg/mL concentration for optimum results. When adding TLR agonists and Golgi-plug, it is essential to dilute these reagents in RPMI-10 and add 100 μL (of 2× concentration) to 100 μL of cell suspension in the 96-well round-bottom plate. Alternatively, a 200 μL of RPMI-10 media with TLR agonists and Golgi-plug (both at 1× concentration) can be added to the wells.
8. Intracellular cytokine staining (ICS) should be carried out using anti-mouse antibodies conjugated with APC or PE or PerCPcy5.5 dyes for better results and use other dyes accordingly for cell surface staining. For ICS staining, cells should be incubated for 25–30 min for better results. Keep cells on ice throughout cell surface and intracellular staining.
9. After surface staining, the cells should be treated with cytofix for 15 min followed by washing and resuspension in FACS buffer. There is no need to add cytofix to ICS-stained cells as these cells were already treated with Cytofix/Cytoperm buffer immediately before ICS staining.
10. For best results, cells should be acquired in flow cytometer within 1–2 days post-staining.

References

1. Shi C, Pamer EG (2011) Monocyte recruitment during infection and inflammation. Nat Rev Immunol 11(11):762–774. https://doi.org/10.1038/nri3070
2. Camp JV, Jonsson CB (2017) A role for neutrophils in viral respiratory disease. Front Immunol 8:550. https://doi.org/10.3389/fimmu.2017.00550
3. Hayashi F, Means TK, Luster AD (2003) Toll-like receptors stimulate human neutrophil function. Blood 102(7):2660–2669. https://doi.org/10.1182/blood-2003-04-1078
4. Prince LR, Whyte MK, Sabroe I, Parker LC (2011) The role of TLRs in neutrophil activation. Curr Opin Pharmacol 11(4):397–403. https://doi.org/10.1016/j.coph.2011.06.007
5. Malmgaard L, Melchjorsen J, Bowie AG, Mogensen SC, Paludan SR (2004) Viral activation of macrophages through TLR-dependent and -independent pathways. J Immunol 173(11):6890–6898
6. Hashimoto Y, Moki T, Takizawa T, Shiratsuchi A, Nakanishi Y (2007) Evidence for phagocytosis of influenza virus-infected, apoptotic cells by neutrophils and macrophages in mice. J Immunol 178(4):2448–2457
7. Gralinski LE, Sheahan TP, Morrison TE, Menachery VD, Jensen K, Leist SR, Whitmore A, Heise MT, Baric RS (2018) Complement activation contributes to severe acute respiratory syndrome coronavirus pathogenesis. MBio 9(5). https://doi.org/10.1128/mBio.01753-18
8. Channappanavar R, Fehr AR, Vijay R, Mack M, Zhao J, Meyerholz DK, Perlman S (2016) Dysregulated type I interferon and inflammatory monocyte-macrophage responses cause lethal pneumonia in SARS-CoV-infected mice. Cell Host Microbe 19(2):181–193. https://doi.org/10.1016/j.chom.2016.01.007
9. Davidson S, Crotta S, McCabe TM, Wack A (2014) Pathogenic potential of interferon alphabeta in acute influenza infection. Nat Commun 5:3864. https://doi.org/10.1038/ncomms4864
10. Tate MD, Deng YM, Jones JE, Anderson GP, Brooks AG, Reading PC (2009) Neutrophils ameliorate lung injury and the development of severe disease during influenza infection. J Immunol 183(11):7441–7450. https://doi.org/10.4049/jimmunol.0902497
11. Brandes M, Klauschen F, Kuchen S, Germain RN (2013) A systems analysis identifies a feed-forward inflammatory circuit leading to lethal influenza infection. Cell 154(1):197–212. https://doi.org/10.1016/j.cell.2013.06.013
12. Hogner K, Wolff T, Pleschka S, Plog S, Gruber AD, Kalinke U, Walmrath HD, Bodner J, Gattenlohner S, Lewe-Schlosser P, Matrosovich M, Seeger W, Lohmeyer J, Herold S (2013) Macrophage-expressed IFN-beta contributes to apoptotic alveolar epithelial cell injury in severe influenza virus pneumonia. PLoS Pathog 9(2):e1003188. https://doi.org/10.1371/journal.ppat.1003188
13. Fujikura D, Chiba S, Muramatsu D, Kazumata M, Nakayama Y, Kawai T, Akira S, Kida H, Miyazaki T (2013) Type-I interferon is critical for FasL expression on lung cells to determine the severity of influenza. PLoS One 8(2):e55321. https://doi.org/10.1371/journal.pone.0055321
14. Teijaro JR, Walsh KB, Cahalan S, Fremgen DM, Roberts E, Scott F, Martinborough E, Peach R, Oldstone MB, Rosen H (2011) Endothelial cells are central orchestrators of cytokine amplification during influenza virus infection. Cell 146(6):980–991. https://doi.org/10.1016/j.cell.2011.08.015

Chapter 16

Histopathologic Evaluation and Scoring of Viral Lung Infection

David K. Meyerholz and Amanda P. Beck

Abstract

Emergent coronaviruses such as MERS-CoV and SARS-CoV can cause significant morbidity and mortality in infected individuals. Lung infection is a common clinical feature and contributes to disease severity as well as viral transmission. Animal models are often required to study viral infections and therapies, especially during an initial outbreak. Histopathology studies allow for identification of lesions and affected cell types to better understand viral pathogenesis and clarify effective therapies. Use of immunostaining allows detection of presumed viral receptors and viral tropism for cells can be evaluated to correlate with lesions. In the lung, lesions and immunostaining can be qualitatively described to define the cell types, microanatomic location, and type of changes seen. These features are important and necessary, but this approach can have limitations when comparing treatment groups. Semiquantitative and quantitative tissue scores are more rigorous as these provide the ability to statistically compare groups and increase the reproducibility and rigor of the study. This review describes principles, approaches, and resources that can be useful to evaluate coronavirus lung infection, focusing on MER-CoV infection as the principal example.

Key words MERS-CoV infection, Lung, Scoring, Pathology, Immunostaining

1 Introduction

Emergent coronaviruses such as severe acute respiratory syndrome (SARS-CoV) and Middle East respiratory syndrome (MERS-CoV) have caused significant impacts on human health, especially during their initial outbreaks [1, 2]. People infected with these coronaviruses often have significant lung disease that contributes to clinical morbidity and mortality [3–5]. Histopathologic examination and immunostaining (e.g., immunohistochemistry) of lung tissues are essential to better understand disease pathogenesis and evaluate novel treatments of these current (and future) virus outbreaks [6–10]. Here, we will focus on MERS-CoV infection to present important principles for valid qualitative and quantitative evaluation of infected lung tissues.

Rahul Vijay (ed.), *MERS Coronavirus: Methods and Protocols*, Methods in Molecular Biology, vol. 2099,
https://doi.org/10.1007/978-1-0716-0211-9_16,

1.1 Factors that Influence Evaluation

Preparation of quality lung tissue samples is important for histopathologic examination to optimize preservation of fine pulmonary architecture and, in the case of immunostaining, antigenicity of target epitopes [11–13]. A study by Engel and Moore identified more than 60 variables in this time frame, beginning with proper sample collection and handling and including multiple aspects of tissue collection, fixation, processing, embedding, slide drying, and storage [14]. Thus, attention to details and quality early will greatly aid the subsequent evaluation, interpretation, and impact of tissue examination.

To collect lungs for histology, samples should be harvested as soon as possible following death to minimize autolysis [11]. Autolysis ("self-digestion") is a postmortem change characterized by degradation of cellular constituents (DNA, RNA, protein) and dissolution of the tissue [15]. Not only can this cause degradation of epitopes and increased nonspecific staining with immunohistochemistry, autolytic regions can be morphologically confused with foci of necrosis and edema [15–17]. If animals will be euthanized, it is preferable to select a method that does not target the lungs such as an intravenous agent. Even use of inhalational overdose of carbon dioxide, as is commonly used in rodents, can potentially cause minor edema/hemorrhage [11, 18, 19]. Evaluation of controls should be standard to evaluate for antemortem or euthanasia-related variables affecting lung evaluation. When examining rodents versus lungs from larger animals or humans, sampling becomes a relevant variable. For instance, mice have small lungs that can be sectioned onto one glass slide for widespread evaluation. Larger sized lungs cannot be sampled adequately using only one slide without introducing sampling bias. Therefore, several samples will need to be collected in larger lungs. The collection method will need to be defined in the methods of publications and should include collection site (standardized vs. lesions sites) and total number, the latter of which depends on the size of the lungs, distribution of lung lesions, and overarching goals of the study.

Proper and adequate fixation of the tissues is essential to retain optimal tissue morphology and cellular antigenicity for immunostaining techniques [11, 20]. However, it is important to remember that if lungs are to be assessed or scored for macroscopic (gross) indicators of disease (such as color, surface texture, and consistency), this must be done prior to fixation, which will affect all of these parameters. Macroscopic evaluation and scoring can be a nice tool to complement histopathology lesions [21, 22]. For tissues that will be paraffin embedded, sections are typically fixed in 10% neutral buffered formalin or 4% paraformaldehyde, though other fixatives may be employed based on the desired analysis endpoints. Collected lung samples can be placed in a minimum of 20:1 volume of fixative:tissue with a maximal thickness of tissue of no more than ~5 mm in at least one dimension to be consistently fixed [20]. For rodents, inflation of the lungs via intratracheal instillation of fixative is recommended to best preserve lung morphology and reduce artifactual atelectasis [15]. However, this approach is contraindicated for lung infection as this can alter the anatomic location of inflammation and cellular debris [22]. The lungs and heart of rodents can be removed en bloc for fixation. Freezing of tissue may be an alternative approach to preserve specific antigens, but this process typically results in suboptimal retention of cellular and tissue architectural detail [11].

After processing to dehydrate the fixed lungs, samples must be embedded and sectioned in a consistent manner. Due to the relatively small size of mice, all lung lobes can be embedded en bloc with the ventral lobar surfaces oriented down in the cassette, which results in sections showing longitudinal views of major conducting airways. An alternative approach for mice, or standard approach for larger species, lung lobes can be collected as multiple sections that fit into a cassette, with each sample embedded separately [11]. Slides are typically stained with hematoxylin and eosin (HE) for routine histologic evaluation. If immunostaining is desired, it is essential to optimize and validate each new antibody utilizing appropriate positive and negative controls to ensure accurate staining results [20, 23]. Similarly, if special histochemical stains will be employed, appropriate control slides and tissues should also be utilized for each batch.

Awareness of normal anatomy and morphology is necessary to recognize any type of change and when utilizing animal models of human disease, this includes knowing differences between the species [24, 25]. For example, there are a number of morphologic differences between the respiratory tract structures of mice and humans. Lobation is distinct, in that mice have four right lung lobes (cranial, middle, caudal, and accessory) and only one left lobe, while humans have three right lung lobes (upper, middle, and lower) and two left lobes (upper and lower) [15, 22]. Rats and mice lack intralobular septa, intrapulmonary bronchi, intrapulmonary submucosal glands, and respiratory bronchioles. Mice also have more club cells extending to the trachea, a thinner blood-gas barrier, and a smaller alveolar diameter than humans [11, 26]. These anatomic variations do not mean that rodents cannot be very valuable models of lung disease; rather they are highlighted here as an example of the type of knowledge necessary for correct interpretation of experimental models.

Inclusion of experienced board-certified pathologists, who are specially trained to examine and interpret tissues changes, as part of the multidisciplinary team can greatly enhance the quality of tissue evaluation [22, 27]. By histopathology, a skilled eye (ideally a pathologist familiar with the model) can not only define the types of inflammatory processes, but also corroborate these findings to clinical signs and/or data from other analyses [22, 27–30]. In addition, pathologists have knowledge of correct lesion nomenclature, as well as potential effects of such variables as strain-related background lesions, husbandry, the microbiome, and diet on the interpretation of results [25]. If pathologists are not involved in designing translational experiments and interpreting lesions in animal models, bias may be introduced and the accuracy of the data and conclusions may be questionable. This approach, which lacks the expertise of a pathologist trained in tissue interpretation, has been labeled as "do-it-yourself pathology" and is linked to multiple publications containing erroneous interpretations [22, 25, 31, 32]. While observations made by biomedical personnel may be biologically accurate in some cases, it is important to note that tissue examination by non-pathologists (even those who are "scientific experts" for a particular disease) is prone to false-positive and false-negative errors and not recommended [33]. Ideally, tissues should be examined by a pathologist familiar with histopathology of the model (*see* **Note 1**). It is recognized that not all labs have access to pathologists for this role and in many situations a member of the investigational team is assigned to the role. In these situations, if possible, it helps to have a pathologist review the study findings prior to publication or have the examiner meet with a pathologist to screen the slides and data for accuracy.

Lungs have unique features compared to other organs that are important for consideration in designing experiments or when making interpretations. For study of infectious diseases, distribution and histologic appearance of lung lesions depends on a variety of factors including the viral inoculate concentration, route of exposure, regional deposition, cellular uptake, chronicity, and host immune response. For instance, inbred mouse strains can have variably sized airways that may affect viral droplet delivery or clinical disease manifestations such as airway obstruction [34]. Inbred mouse strains can also exhibit biased (e.g., Th1 vs. Th2 immune responses) or deficient immune signaling pathways that might influence infection susceptibility or severity [35, 36]. Sex can also be an influencing factor for infection and needs to be considered in the experimental design [37]. Even actions as simple as laying an animal in lateral recumbency to recover from anesthesia following viral inoculation may lead to more prominent lesions in certain lobe(s) [22]. For many of these features, inclusion of appropriate control animals (i.e., strain-, age-, and sex-matched, housed under identical husbandry conditions and free from confounding pathogens) is necessary and important to tease out any lesions unrelated to the treatments. Unlike the other organs in which the size is relatively static, the lung has dynamic size changes during normal respiration. Handing of the postmortem lung in a standardized manner is useful to prevent postmortem atelectasis or variable inter-animal insufflation. Right ventricular perfusion of fixative into the lungs prior to extraction can help with fixation as well as insufflate the airspaces without dislodging inflammation or mucocellular debris [22].

1.2 Histopathology

Histopathology is the microscopic examination of tissues for morphologic or structural changes that differ from normal and these changes are called lesions. Histopathology of coronavirus-infected lung in humans and animal models can be a useful tool to help define affected cells, illuminate the structural cause(s) of clinical signs, and clarify potential therapies. During disease outbreaks, clinical data including autopsy cases can be studied in parallel with animal model investigations to better define lung disease pathogenesis and therapies. For instance, in 2012 the novel human coronavirus known as MERS-CoV was first isolated from a patient dying in Saudi Arabia [2, 38]. In the region of the outbreak, local burial rituals along with the requirement for high biosecurity constrained autopsy studies from being performed until the first report in early 2016 [4]. Within a few years of the first reported MERS-CoV case in humans in 2012, several animal models were being studied and these models provided much of the initial critically important lung pathology data [39–44].

Histopathologic examination of viral lung infection requires awareness of any anticipated lesions from clinical or published data, as it is available. Examples of MERS-CoV lesions are listed in Table 1. For instance, acute diffuse alveolar damage (DAD) is a common feature of MERS-CoV lung lesions and it is composed of lesions such as edema, inflammation, and alveolar septal injury [4, 48–50]. While awareness of reported lesions can help guide the pathologist in examination, it is also useful to have a consistent method for examination of experimental tissues to avoid unintentional bias that might cause a failure to detection of unexpected lesions [51]. Consistent examination of all tissues from control and treatment groups can reduce the chances of mistakenly diagnosing nonspecific model background phenotype as a MERS-CoV-specific

Table 1
Examples of lesions seen in MERS-CoV lung infections

Lesions	Necrosis/cell death [45]
	Edema [8, 21, 45]
	Hyaline membranes/fibrin [21]
	Inflammation [8]
	Thrombi [8, 46]
	Congestion [8]
	Hemorrhage [45, 46]
	Pneumonia [46, 47]
	Type II hyperplasia [47]
	Syncytia [47]

Table 2
Methods of masking to prevent observer bias [22, 52–54]

Method	Approach	Usage
Comprehensive	Samples are labeled without group identification (1, 2, 3, 4 ...), minimal background information provided	Allows for experienced observers to score well-defined models, otherwise susceptible to errors
Grouped	Samples are labeled according to de-identified groups (A1, A2, A3, B1, B2, B3...)	Allows for masked evaluation of groups while observer is informed about experimental context
Post-examination	Samples are examined in a transparent manner to determine the type and scope of tissue changes, samples are then masked for scoring	Allows for full examination and disclosure of experimental context; groups with small N may let observer recall sample group assignment

lesion [22, 25, 29]. For instance, a lesion that is present in the controls and treatment groups can be defined as a background model/technique phenotype and should not be reported as a MERS-CoV specific lesion. Masking of the pathologist to the group assignments is useful to avoid observer bias and each type of masking method has certain advantages and limitations (Table 2, *see* **Note 2**) [22]. A common approach for histopathologic examination is to start at low magnification to screen for any obvious lesions and assess quality of the tissue section (*see* **Note 3**). This allows examination of microscopic structures such as airways, alveoli, alveolar septa, air spaces, vessels, and pleura. Examination at high magnification allows for screening of cellular and interstitial components of each structure for lesions (e.g., injury, inflammation, necrosis). Most slides will be examined using HE, but additional stains can be used on serial sections to further define any changes. For instance, mucus in goblet cells or secreted into air spaces can be highlighted by special stains like Periodic acid Schiff in glycogen-depleted tissues or Alcian blue [55, 56].

After the slides and stains have been examined for all groups, the results will need to be prepared for publication. Qualitative characterization of the findings is very important to understand features of the disease including cellular tropism, anatomic predisposition, and nature of lesions leading to clinical signs (*see* **Note 4**) [10, 21]. Qualitative descriptions of lesions include type (e.g., epithelial sloughing/necrosis), location (e.g., alveoli), distribution (e.g., locally extensive), inflammation (e.g., neutrophilic), and cell types involved (e.g., type I pneumocytes). Qualitative features can be sufficiently described in the text and exemplified in representative figures. Use of arrows and other forms of annotation are valuable in figures to clarify and guide readers through the images. High-quality descriptions will help the reader (including reviewers) better understand what was seen and allow for others to reproduce the study.

1.3 Immunostaining

Immunostaining (immunohistochemistry) is a valuable tool in viral lung disease investigations as it can be used to study cellular localization of receptors and viral targets. For instance, detection of the MERS-CoV receptor dipeptidyl peptidase 4 (DPP4) virus receptor can give insights to cell tropism to help explain disease pathogenesis [4, 6, 21, 42, 57–59].

There are several tissue handling (preanalytical) factors that can significantly affect the quality and specificity of immunostaining and its analysis. These have been discussed earlier sections of the paper and in several reviews [20, 52, 60–63]. Similarly, there are many factors during the staining procedure that itself can also influence the results. Deparaffinization, lack of control tissues, optimization/validation techniques, species, batch effects, and chromogens can all influence the final quality and assessment of immunostaining methods. Standard operating procedures for each of the technical steps, if used by all biomedical staff, can significantly mitigate many of these issues. Use of positive and negative control tissues for each batch of immunostained tissues can help in validating appropriate staining and also making clear any potential nonspecific immunostaining. After the stained slides have been examined for all groups, qualitative statements about the immunostaining can be made and prepared for publication text and images. Descriptive text of immunostaining (receptor or virus) could include cell types (e.g., type I pneumocytes), cell integrity (necrotic vs. intact cells), and subcellular location (e.g., diffuse cytoplasmic). Demonstration of immunostaining using annotated images can strengthen the qualitative data.

1.4 Scoring

As shown above, qualitative descriptions of tissue changes are useful and necessary, but they are less applicable in terms of group comparisons. More robust and reproducible methods are desirable and these criteria can be sought in tissue scoring systems (semiquantitative and quantitative) that produce data that allow for statistical analyses for evaluation of group differences (*see* **Note 5**) [52, 53]. Importantly, these scoring principles can be applied to tissue lesions (gross and/or histopathologic) as well as immunostained sections.

1.4.1 Nominal Approaches

Nominal approaches do not score or make quantitative measurements on tissue samples, but rather each sample is assigned to well-defined categories [52, 54]. The numbers of samples assigned to each category are recorded and evaluated with appropriate statistical tests. As a simple mock example, consider examining the lungs of wild-type (WT) or mutated mice for the presence or absence of edema, a common feature of DAD. Each mouse would be assigned to either “no edema” (Fig. 1a) or “edema” (Fig. 1b, c) categories. If 10 mice per group were evaluated, the WT group might have nine with edema and one without, while the mutated group has

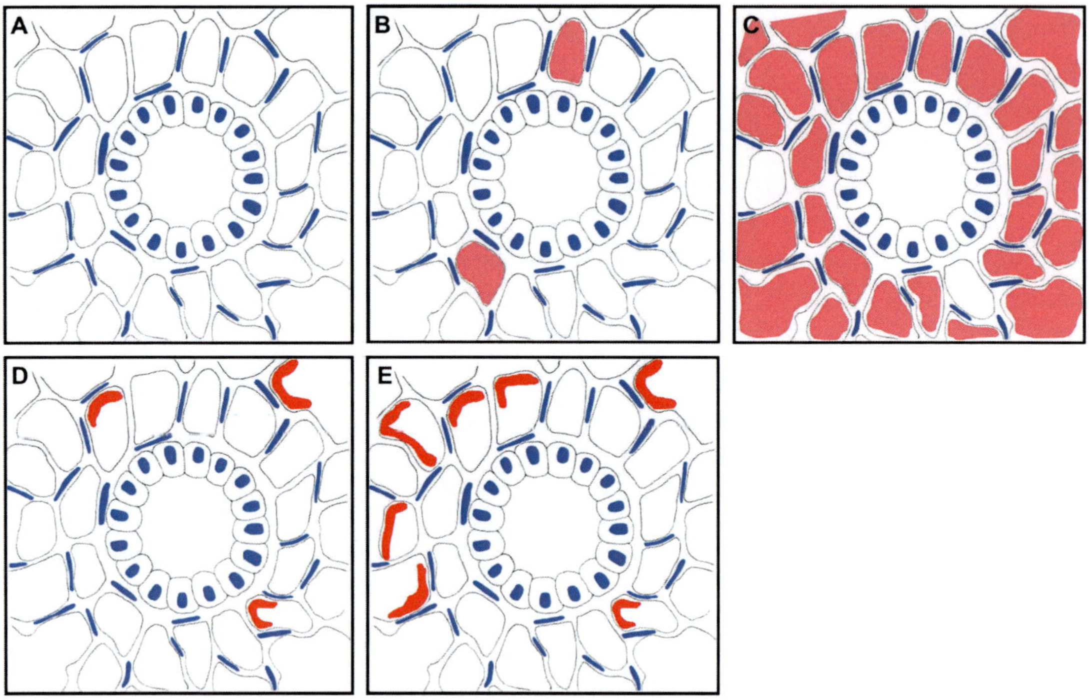

Fig. 1 Mock example of mouse lung lesions during MERS-CoV infection. (**a**) Normal bronchiole and alveolar structures. (**b**, **c**) Pulmonary edema (pink color filling alveoli). (**d**, **e**) Hyaline membranes (red crescents lining alveolar walls)

three with edema and seven without. Evaluating these data using a Fisher exact test results in a significant difference ($P = 0.02$) between WT and mutated mice. The presence of any lesion or immunostaining can be similarly assessed in this manner, but it is important to have clear guidelines or thresholds to distinguish the categories.

1.4.2 Semiquantitative Approaches

Semiquantitative approaches are used to transform qualitative tissue changes into numerical scores using specific morphologic criteria [52, 53]. Semiquantitative methods have several advantages in that they can be done with minimal technical resources, quickly at the microscope for small to medium studies, provide guidance for future quantitative studies, and provide complementary data for publication [52–54]. The most commonly used semiquantitative methods produce ordinal scores. Ordinal implies there is an order or progression of severity in the assigned grades that define each score, with typically four to five grades being optimal (e.g., 0, 1, 2, 3, 4). Each grade should be well defined so there is minimal ambiguity in assigning samples. Use of simple descriptive modifiers such as normal, rare, mild, moderate, and severe is discouraged as these have different meanings for each observer and thus limit reproducibility of the scoring. As a mock example of ordinal scoring, WT and mutated mice might be evaluated for the extent of hyaline membranes lining alveolar walls. The scoring grades might look like: "0"—none, "1"— <25% (*see* Fig. 1d), "2"—26–50% (*see* Fig. 1e), "3"—51–75%, and "4" >75% of alveolar walls in the lung section. If the ordinal scoring for seven mice per group produced the following results for WT (3, 3, 2, 3, 4, 3, 4) and mutated mice (1, 2, 1, 1, 1, 1, 2), then the data can be statistically analyzed. Importantly, ordinal scores do not meet the assumptions required for parametric tests; thus nonparametric tests should be used [33]. For the mock example, the difference between groups using a Mann-Whitney U-test was significant ($P = 0.002$).

1.4.3 Quantitative Approaches

Quantitative methods are tissue techniques that measure specific tissue components (length, area, volume, number, percentage, etc.) [52]. Quantitative methods tend to have greater precision and sensitivity than semiquantitative methods. These methods often require high-quality images and specialized software to properly analyze the tissues, which can make the methods costlier for some labs than semiquantitative techniques. The growing interest in automation and artificial intelligence may increase future efficiency and cost-effectiveness of quantification of tissue parameters, especially for large projects [64–67].

Quantification of viral lesions and immunostaining in tissues is an option; however, quantification is not commonly performed in tissue sections due to potential confounding factors such as random distribution of viral inoculum and difficulty in objectively quantifying lesions. If choosing to perform quantitative scoring, evaluation of clinically relevant anatomic compartments (airways or alveoli) can help standardize the assessment. As a mock example, viral immunostaining could be evaluated as a percent of cell number in mouse bronchioles (Fig. 2a–c; 0%, 12.5%, and 43.8%, respectively) or as an alternative one could also assess the area of immunostaining as a percent of the bronchiolar epithelium area. In contrast, the alveolar compartment can be more difficult to assess than airways because of their thin walls, which makes evidence of necrosis/sloughing or immunostaining a challenge. To normalize analysis, one could assess the percent of alveoli with immunostaining (Fig. 2d–e). However, this would likely require extensive time/labor or specialized software. If quantitation is not feasible but is an important variable, one could revert to semiquantitative scoring to assess immunostaining as a percentage of affected alveolar walls. Using the distribution scoring system defined for Fig. 1, one could score the samples in Fig. 2d–e, as ordinal scores of 1 and 4, respectively. While the mock example is simple, reality often paints a more complex portrait of lesion or immunostaining distribution (Fig. 2f).

When it comes to tissue scoring, each project is unique. Investigators will have to evaluate the lung samples to determine the best scoring approaches in relation to the breadth of lesions and goals of

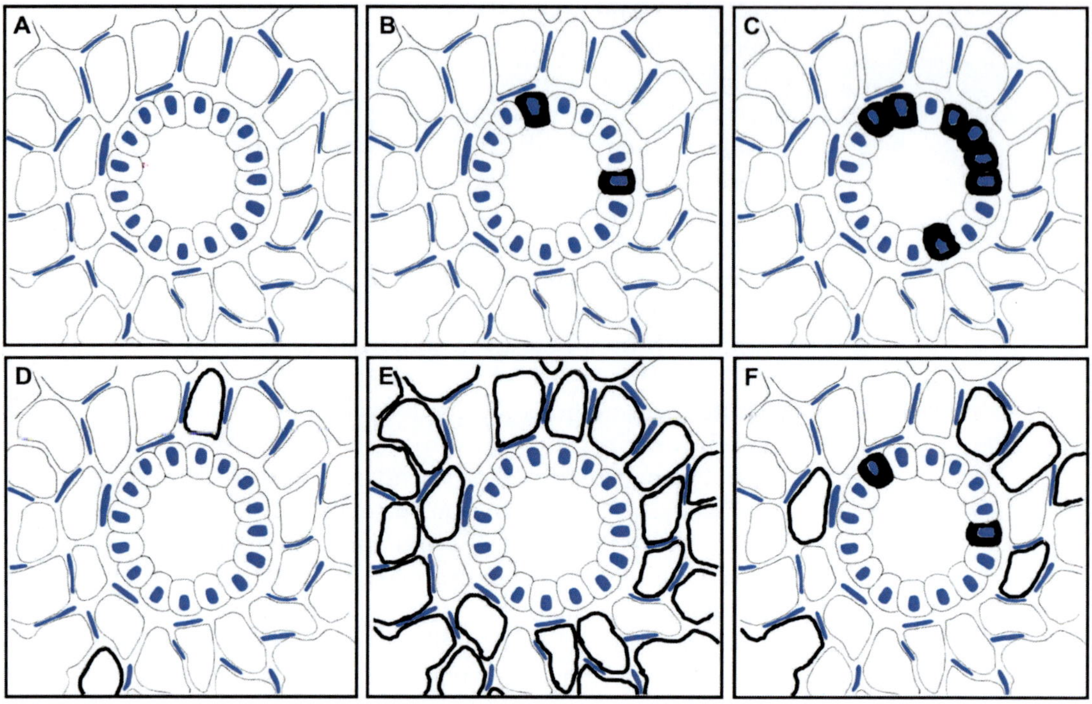

Fig. 2 Mock example of viral immunostaining during MERS-CoV infection. (**a**) No immunostaining in control lung. (**b**, **c**) Immunostaining (black color) in airways. (**d**, **e**) Immunostaining in alveoli. (**f**) Immunostaining in airway and alveoli

the project. Most importantly, any scoring that is performed should be corroborated, when possible, with other data to validate the findings [22, 52, 53]. For instance, if group A has more immunostaining than group B, this could be validated by ELISA or Western blots of whole lung homogenates. Alternatively, lesion severity could be corroborated to measurements of clinical data (*see* **Note 6**). Validation can help give more confidence in the data rigor and reproducibility.

1.4.4 Statistical Analyses

Inappropriate use of paired t-tests and shopping for significance are two issues that have slipped into the published literature and potentially compromise the interpretation and reproducibility of studies [33]. For the various scoring methods, statistical analyses of the data should involve the collaborative expertise of a statistician to be able to identify the most relevant tests to confidently evaluate for group differences [22, 33, 52, 53].

2 Summary

Examination of infected lung tissues for histopathology and immunostaining are common and needed approaches to study viral lung infection, especially in emergent coronaviruses like MERS-CoV.

Following the principles and concepts above will help guide and lead studies to more valid and reproducible data.

3 Notes

1. Ideally a pathologist familiar with the model is available for the lab to evaluate experimental tissues. If not, then a pathologist collaborator should be sought to perform or review of the results of examination prior to submission for publication. This prevents publication of data that is flawed or needs subsequent retraction.
2. Masking is important to prevent potential bias by the observer pathologist (Table 2). For new projects, the post-examination is preferred as this helps the pathologist understand the goals/experimental design of the project as well as see quality and scope of lesions/stains. For most other research projects where the pathologist is familiar with the model, these can be masked in grouped fashion to maximize the interpretative power of the pathologist to screen for biologically relevant changes in a group-specific manner. Comprehensive masking is often discouraged as it effectively constrains the ability of the pathologist in defining relevant versus unconnected data and therefore limits the sensitivity and specificity of the pathology data.
3. Evaluation of slides from all treatment and control groups prior to detailed examination is useful to give the pathologist an overview and primer of the type, scope and severity of lesions/stains.
4. Detailed examination of the tissues allows for extrapolation of qualitative descriptive data. If there are questions regarding the cells/tissues that can be addressed by specific stains—these could be done at this time to corroborate/clarify descriptive findings.
5. When biologically relevant lesions are defined in the project, group-specific changes may be evaluated for by semiquantitative or quantitative scores. Semiquantitative approaches are often done initially and the results can be used as screening tools to set up primary scoring approaches or be used as primary/supplemental data for reporting group differences in lesions or stains. Quantitative approaches may be performed by at the microscope (e.g., cell counts) or automated on digital images by specialized software.

   Regardless of the masking method (*see* **Note 2**), it is often useful to score the slides in a random masked fashion and in one sitting to prevent diagnostic drift. After scoring, it is sometimes beneficial to take scoring data to see if these same differences are morphologically detectable in the respective groups. If the pathologist can see these differences, it gives further confidence to the scoring approach and final interpretations. If not, it can raise questions as to the scoring methods.

6. Effective reporting of pathology data requires transparency of methods, numbers of animals, statistical analyses, etc. Producing graphs of scoring data with matching images that are annotated can be very powerful tools in conveying the results to readers.

Acknowledgements

We would like to thank The Lamb Art Studio (www.faithlamb.com) for excellent assistance with graphic artwork.

References

1. Booth CM, Matukas LM, Tomlinson GA, Rachlis AR, Rose DB, Dwosh HA, Walmsley SL, Mazzulli T, Avendano M, Derkach P, Ephtimios IE, Kitai I, Mederski BD, Shadowitz SB, Gold WL, Hawryluck LA, Rea E, Chenkin JS, Cescon DW, Poutanen SM, Detsky AS (2003) Clinical features and short-term outcomes of 144 patients with SARS in the greater Toronto area. JAMA 289(21):2801–2809. https://doi.org/10.1001/jama.289.21.JOC30885
2. Zumla A, Hui DS, Perlman S (2015) Middle East respiratory syndrome. Lancet 386 (9997):995–1007. https://doi.org/10.1016/S0140-6736(15)60454-8
3. Khalid I, Alraddadi BM, Dairi Y, Khalid TJ, Kadri M, Alshukairi AN, Qushmaq IA (2016) Acute management and long-term survival among subjects with severe Middle East respiratory syndrome coronavirus pneumonia and ARDS. Respir Care 61(3):340–348. https://doi.org/10.4187/respcare.04325
4. Ng DL, Al Hosani F, Keating MK, Gerber SI, Jones TL, Metcalfe MG, Tong S, Tao Y, Alami NN, Haynes LM, Mutei MA, Abdel-Wareth L, Uyeki TM, Swerdlow DL, Barakat M, Zaki SR (2016) Clinicopathologic, immunohistochemical, and ultrastructural findings of a fatal case of Middle East respiratory syndrome coronavirus infection in the United Arab Emirates, april 2014. Am J Pathol 186(3):652–658. https://doi.org/10.1016/j.ajpath.2015.10.024
5. Peiris JS, Chu CM, Cheng VC, Chan KS, Hung IF, Poon LL, Law KI, Tang BS, Hon TY, Chan CS, Chan KH, Ng JS, Zheng BJ, Ng WL, Lai RW, Guan Y, Yuen KY, Group HUSS (2003) Clinical progression and viral load in a community outbreak of coronavirus-associated SARS pneumonia: a prospective study. Lancet 361(9371):1767–1772
6. Cockrell AS, Johnson JC, Moore IN, Liu DX, Bock KW, Douglas MG, Graham RL, Solomon J, Torzewski L, Bartos C, Hart R, Baric RS, Johnson RF (2018) A spike-modified Middle East respiratory syndrome coronavirus (MERS-CoV) infectious clone elicits mild respiratory disease in infected rhesus macaques. Sci Rep 8(1):10727. https://doi.org/10.1038/s41598-018-28900-1
7. Hua X, Vijay R, Channappanavar R, Athmer J, Meyerholz DK, Pagedar N, Tilley S, Perlman S (2018) Nasal priming by a murine coronavirus provides protective immunity against lethal heterologous virus pneumonia. JCI Insight 3 (11):99025. https://doi.org/10.1172/jci.insight.99025
8. Li K, Wohlford-Lenane C, Perlman S, Zhao J, Jewell AK, Reznikov LR, Gibson-Corley KN, Meyerholz DK, McCray PB Jr (2016) Middle East respiratory syndrome coronavirus causes multiple organ damage and lethal disease in mice transgenic for human dipeptidyl peptidase 4. J Infect Dis 213(5):712–722. https://doi.org/10.1093/infdis/jiv499
9. Menachery VD, Yount BL Jr, Debbink K, Agnihothram S, Gralinski LE, Plante JA, Graham RL, Scobey T, Ge XY, Donaldson EF, Randell SH, Lanzavecchia A, Marasco WA, Shi ZL, Baric RS (2015) A SARS-like cluster of circulating bat coronaviruses shows potential for human emergence. Nat Med 21 (12):1508–1513. https://doi.org/10.1038/nm.3985

10. Meyerholz DK, Lambertz AM, McCray PB Jr (2016) Dipeptidyl peptidase 4 distribution in the human respiratory tract: implications for the Middle East respiratory syndrome. Am J Pathol 186(1):78–86. https://doi.org/10.1016/j.ajpath.2015.09.014

11. Aeffner F, Bolon B, Davis IC (2015) Mouse models of acute respiratory distress syndrome: a review of analytical approaches, pathologic features, and common measurements. Toxicol Pathol 43(8):1074–1092. https://doi.org/10.1177/0192623315598399

12. Dunstan RW, Wharton KA Jr, Quigley C, Lowe A (2011) The use of immunohistochemistry for biomarker assessment-can it compete with other technologies? Toxicol Pathol 39 (6):988–1002. https://doi.org/10.1177/0192623311419163

13. Eaton KA, Danon SJ, Krakowka S, Weisbrode SE (2007) A reproducible scoring system for quantification of histologic lesions of inflammatory disease in mouse gastric epithelium. Comp Med 57(1):57–65

14. Engel KB, Moore HM (2011) Effects of preanalytical variables on the detection of proteins by immunohistochemistry in formalin-fixed, paraffin-embedded tissue. Arch Pathol Lab Med 135(5):537–543. https://doi.org/10.1043/2010-0702-RAIR.1

15. Renne R, Brix A, Harkema J, Herbert R, Kittel B, Lewis D, March T, Nagano K, Pino M, Rittinghausen S, Rosenbruch M, Tellier P, Wohrmann T (2009) Proliferative and nonproliferative lesions of the rat and mouse respiratory tract. Toxicol Pathol 37 (7 Suppl):5S–73S. https://doi.org/10.1177/0192623309353423

16. Rollins KE, Meyerholz DK, Johnson GD, Capparella AP, Loew SS (2012) A forensic investigation into the etiology of bat mortality at a wind farm: barotrauma or traumatic injury? Vet Pathol 49(2):362–371. https://doi.org/10.1177/0300985812436745

17. Williams JM, Duckworth CA, Burkitt MD, Watson AJ, Campbell BJ, Pritchard DM (2015) Epithelial cell shedding and barrier function: a matter of life and death at the small intestinal villus tip. Vet Pathol 52 (3):445–455. https://doi.org/10.1177/0300985814559404

18. Boivin GP, Bottomley MA, Schiml PA, Goss L, Grobe N (2017) Physiologic, behavioral, and histologic responses to various euthanasia methods in C57BL/6NTac male mice. J Am Assoc Lab Anim Sci 56(1):69–78

19. Marquardt N, Feja M, Hunigen H, Plendl J, Menken L, Fink H, Bert B (2018) Euthanasia of laboratory mice: are isoflurane and sevoflurane real alternatives to carbon dioxide? PLoS One 13(9):e0203793. https://doi.org/10.1371/journal.pone.0203793

20. Olivier AK, Naumann P, Goeken A, Hochstedler C, Sturm M, Rodgers JR, Gibson-Corley KN, Meyerholz DK (2012) Genetically modified species in research: opportunities and challenges for the histology core laboratory. J Histotechnol 35(2):63–67

21. Li K, Wohlford-Lenane CL, Channappanavar R, Park JE, Earnest JT, Bair TB, Bates AM, Brogden KA, Flaherty HA, Gallagher T, Meyerholz DK, Perlman S, McCray PB Jr (2017) Mouse-adapted MERS coronavirus causes lethal lung disease in human DPP4 knockin mice. Proc Natl Acad Sci U S A 114(15):E3119–E3128. https://doi.org/10.1073/pnas.1619109114

22. Meyerholz DK, Sieren JC, Beck AP, Flaherty HA (2018) Approaches to evaluate lung inflammation in translational research. Vet Pathol 55(1):42–52. https://doi.org/10.1177/0300985817726117

23. Gibson-Corley KN, Hochstedler C, Sturm M, Rogers J, Olivier AK, Meyerholz DK (2012) Successful integration of the histology core laboratory in translational research. J Histotechnol 35(1):17–21. https://doi.org/10.1179/2046023612Y.0000000001

24. Klopfleisch R (2013) Multiparametric and semiquantitative scoring systems for the evaluation of mouse model histopathology-a systematic review. BMC Vet Res 9:123. https://doi.org/10.1186/1746-6148-9-123

25. Ward JM, Schofield PN, Sundberg JP (2017) Reproducibility of histopathological findings in experimental pathology of the mouse: a sorry tail. Lab Anim (NY) 46(4):146–151. https://doi.org/10.1038/laban.1214

26. Meyerholz DK, Suarez CJ, Dintis SM, Frevert CW (2018) Respiratory system. In: Treuting PM, Dintis SM, Montine KS (eds) Comparative anatomy and histology: a mouse, rat and human atlas, 2nd edn. Elsevier, London, pp 147–162

27. Adissu HA, Estabel J, Sunter D, Tuck E, Hooks Y, Carragher DM, Clarke K, Karp NA, Sanger Mouse Genetics P, Newbigging S, Jones N, Morikawa L, White JK, McKerlie C (2014) Histopathology reveals correlative and unique phenotypes in a high-throughput mouse phenotyping screen. Dis Model Mech 7(5):515–524. https://doi.org/10.1242/dmm.015263

28. Ladiges W, Ikeno Y, Liggitt D, Treuting PM (2013) Pathology is a critical aspect of preclinical aging studies. Pathobiol Aging Age Relat

Dis:3. https://doi.org/10.3402/pba.v3i0.22451

29. Treuting PM, Snyder JM, Ikeno Y, Schofield PN, Ward JM, Sundberg JP (2016) The vital role of pathology in improving reproducibility and translational relevance of aging studies in Rodents. Vet Pathol 53(2):244–249. https://doi.org/10.1177/0300985815620629
30. Ward JM, Youssef SA, Treuting PM (2016) Why animals die: an introduction to the pathology of aging. Vet Pathol 53(2):229–232. https://doi.org/10.1177/0300985815612151
31. Cardiff RD, Ward JM, Barthold SW (2008) 'One medicine-one pathology': are veterinary and human pathology prepared? Lab Investig 88(1):18–26. https://doi.org/10.1038/labinvest.3700695
32. Wolf JC, Wheeler JR (2018) A critical review of histopathological findings associated with endocrine and non-endocrine hepatic toxicity in fish models. Aquat Toxicol 197:60–78. https://doi.org/10.1016/j.aquatox.2018.01.013
33. Meyerholz DK, Tintle NL, Beck AP (2019) Common pitfalls in analysis of tissue scores. Vet Pathol 56(1):39–42. https://doi.org/10.1177/0300985818794250
34. Thiesse J, Namati E, Sieren JC, Smith AR, Reinhardt JM, Hoffman EA, McLennan G (2010) Lung structure phenotype variation in inbred mouse strains revealed through in vivo micro-CT imaging. J Appl Physiol (1985) 109(6):1960–1968. https://doi.org/10.1152/japplphysiol.01322.2009
35. Feng CG, Scanga CA, Collazo-Custodio CM, Cheever AW, Hieny S, Caspar P, Sher A (2003) Mice lacking myeloid differentiation factor 88 display profound defects in host resistance and immune responses to Mycobacterium avium infection not exhibited by Toll-like receptor 2 (TLR2)- and TLR4-deficient animals. J Immunol 171(9):4758–4764. https://doi.org/10.4049/jimmunol.171.9.4758
36. Schulte S, Sukhova GK, Libby P (2008) Genetically programmed biases in Th1 and Th2 immune responses modulate atherogenesis. Am J Pathol 172(6):1500–1508. https://doi.org/10.2353/ajpath.2008.070776
37. Channappanavar R, Fett C, Mack M, Ten Eyck PP, Meyerholz DK, Perlman S (2017) Sex-based differences in susceptibility to severe acute respiratory syndrome coronavirus infection. J Immunol 198(10):4046–4053. https://doi.org/10.4049/jimmunol.1601896
38. Zaki AM, van Boheemen S, Bestebroer TM, Osterhaus AD, Fouchier RA (2012) Isolation of a novel coronavirus from a man with pneumonia in Saudi Arabia. N Engl J Med 367(19):1814–1820. https://doi.org/10.1056/NEJMoa1211721
39. Coleman CM, Matthews KL, Goicochea L, Frieman MB (2014) Wild-type and innate immune-deficient mice are not susceptible to the Middle East respiratory syndrome coronavirus. J Gen Virol 95(Pt 2):408–412. https://doi.org/10.1099/vir.0.060640-0
40. de Wit E, Prescott J, Baseler L, Bushmaker T, Thomas T, Lackemeyer MG, Martellaro C, Milne-Price S, Haddock E, Haagmans BL, Feldmann H, Munster VJ (2013) The Middle East respiratory syndrome coronavirus (MERS-CoV) does not replicate in Syrian hamsters. PLoS One 8(7):e69127. https://doi.org/10.1371/journal.pone.0069127
41. de Wit E, Rasmussen AL, Falzarano D, Bushmaker T, Feldmann F, Brining DL, Fischer ER, Martellaro C, Okumura A, Chang J, Scott D, Benecke AG, Katze MG, Feldmann H, Munster VJ (2013) Middle East respiratory syndrome coronavirus (MERS-CoV) causes transient lower respiratory tract infection in rhesus macaques. Proc Natl Acad Sci U S A 110(41):16598–16603. https://doi.org/10.1073/pnas.1310744110
42. Meyerholz DK (2016) Modeling emergent diseases: lessons from Middle East respiratory syndrome. Vet Pathol 53(3):517–518. https://doi.org/10.1177/0300985816634811
43. Yao Y, Bao L, Deng W, Xu L, Li F, Lv Q, Yu P, Chen T, Xu Y, Zhu H, Yuan J, Gu S, Wei Q, Chen H, Yuen KY, Qin C (2014) An animal model of MERS produced by infection of rhesus macaques with MERS coronavirus. J Infect Dis 209(2):236–242. https://doi.org/10.1093/infdis/jit590
44. Zhao J, Li K, Wohlford-Lenane C, Agnihothram SS, Fett C, Zhao J, Gale MJ Jr, Baric RS, Enjuanes L, Gallagher T, McCray PB Jr, Perlman S (2014) Rapid generation of a mouse model for Middle East respiratory syndrome. Proc Natl Acad Sci U S A 111(13):4970–4975. https://doi.org/10.1073/pnas.1323279111
45. Yu P, Xu Y, Deng W, Bao L, Huang L, Xu Y, Yao Y, Qin C (2017) Comparative pathology of rhesus macaque and common marmoset animal models with Middle East respiratory syndrome coronavirus. PLoS One 12(2):e0172093. https://doi.org/10.1371/journal.pone.0172093
46. Dietert K, Gutbier B, Wienhold SM, Reppe K, Jiang X, Yao L, Chaput C, Naujoks J, Brack M, Kupke A, Peteranderl C, Becker S, von Lachner C, Baal N, Slevogt H, Hocke AC, Witzenrath M, Opitz B, Herold S,

Hackstein H, Sander LE, Suttorp N, Gruber AD (2017) Spectrum of pathogen- and model-specific histopathologies in mouse models of acute pneumonia. PLoS One 12(11): e0188251. https://doi.org/10.1371/journal.pone.0188251

47. Falzarano D, de Wit E, Feldmann F, Rasmussen AL, Okumura A, Peng X, Thomas MJ, van Doremalen N, Haddock E, Nagy L, LaCasse R, Liu T, Zhu J, McLellan JS, Scott DP, Katze MG, Feldmann H, Munster VJ (2014) Infection with MERS-CoV causes lethal pneumonia in the common marmoset. PLoS Pathog 10(8): e1004250. https://doi.org/10.1371/journal.ppat.1004250
48. Alsaad KO, Hajeer AH, Al Balwi M, Al Moaiqel M, Al Oudah N, Al Ajlan A, AlJohani S, Alsolamy S, Gmati GE, Balkhy H, Al-Jahdali HH, Baharoon SA, Arabi YM (2018) Histopathology of Middle East respiratory syndrome coronovirus (MERS-CoV) infection—clinicopathological and ultrastructural study. Histopathology 72(3):516–524. https://doi.org/10.1111/his.13379
49. Fehr AR, Channappanavar R, Perlman S (2017) Middle East respiratory syndrome: emergence of a pathogenic human coronavirus. Annu Rev Med 68:387–399. https://doi.org/10.1146/annurev-med-051215-031152
50. Sweeney RM, McAuley DF (2016) Acute respiratory distress syndrome. Lancet 388 (10058):2416–2430. https://doi.org/10.1016/S0140-6736(16)00578-X
51. Drew T, Vo ML, Wolfe JM (2013) The invisible gorilla strikes again: sustained inattentional blindness in expert observers. Psychol Sci 24 (9):1848–1853. https://doi.org/10.1177/0956797613479386
52. Meyerholz DK, Beck AP (2018) Principles and approaches for reproducible scoring of tissue stains in research. Lab Investig 98 (7):844–855. https://doi.org/10.1038/s41374-018-0057-0
53. Meyerholz DK, Beck AP (2019) Fundamental concepts for semiquantitative tissue scoring in translational research. ILAR J 59:13–17. https://doi.org/10.1093/ilar/ily025
54. Gibson-Corley KN, Olivier AK, Meyerholz DK (2013) Principles for valid histopathologic scoring in research. Vet Pathol 50 (6):1007–1015. https://doi.org/10.1177/0300985813485099
55. Meyerholz DK, Beck AP, Goeken JA, Leidinger MR, Ofori-Amanfo GK, Brown HC, Businga TR, Stoltz DA, Reznikov LR, Flaherty HA (2018) Glycogen depletion can increase the specificity of mucin detection in airway tissues. BMC Res Notes 11(1):763. https://doi.org/10.1186/s13104-018-3855-y
56. Vergara-Alert J, Vidal E, Bensaid A, Segales J (2017) Searching for animal models and potential target species for emerging pathogens: experience gained from Middle East respiratory syndrome (MERS) coronavirus. One Health 3:34–40. https://doi.org/10.1016/j.onehlt.2017.03.001
57. Haagmans BL, van den Brand JM, Provacia LB, Raj VS, Stittelaar KJ, Getu S, de Waal L, Bestebroer TM, van Amerongen G, Verjans GM, Fouchier RA, Smits SL, Kuiken T, Osterhaus AD (2015) Asymptomatic Middle East respiratory syndrome coronavirus infection in rabbits. J Virol 89(11):6131–6135. https://doi.org/10.1128/JVI.00661-15
58. Vergara-Alert J, van den Brand JM, Widagdo W, Munoz MT, Raj S, Schipper D, Solanes D, Cordon I, Bensaid A, Haagmans BL, Segales J (2017) Livestock susceptibility to infection with Middle East respiratory syndrome coronavirus. Emerg Infect Dis 23 (2):232–240. https://doi.org/10.3201/eid2302.161239
59. Widagdo W, Raj VS, Schipper D, Kolijn K, van Leenders G, Bosch BJ, Bensaid A, Segales J, Baumgartner W, Osterhaus A, Koopmans MP, van den Brand JMA, Haagmans BL (2016) Differential expression of the Middle East respiratory syndrome coronavirus receptor in the upper respiratory tracts of humans and dromedary camels. J Virol 90(9):4838–4842. https://doi.org/10.1128/JVI.02994-15
60. Janardhan KS, Jensen H, Clayton NP, Herbert RA (2018) Immunohistochemistry in investigative and toxicologic pathology. Toxicol Pathol 46(5):488–510. https://doi.org/10.1177/0192623318776907
61. Kim SW, Roh J, Park CS (2016) Immunohistochemistry for pathologists: protocols, pitfalls, and tips. J Pathol Transl Med 50 (6):411–418. https://doi.org/10.4132/jptm.2016.08.08
62. Ramos-Vara JA, Miller MA (2014) When tissue antigens and antibodies get along: revisiting the technical aspects of immunohistochemistry-the red, brown, and blue technique. Vet Pathol 51(1):42–87. https://doi.org/10.1177/0300985813505879
63. Ward JM, Rehg JE (2014) Rodent immunohistochemistry: pitfalls and troubleshooting. Vet Pathol 51(1):88–101. https://doi.org/10.1177/0300985813503571
64. Aeffner F, Adissu HA, Boyle MC, Cardiff RD, Hagendorn E, Hoenerhoff MJ, Klopfleisch R, Newbigging S, Schaudien D, Turner O, Wilson

K (2018) Digital microscopy, image analysis, and virtual slide repository. ILAR J 59:66–79. https://doi.org/10.1093/ilar/ily007

65. Barisoni L, Gimpel C, Kain R, Laurinavicius A, Bueno G, Zeng C, Liu Z, Schaefer F, Kretzler M, Holzman LB, Hewitt SM (2017) Digital pathology imaging as a novel platform for standardization and globalization of quantitative nephropathology. Clin Kidney J 10 (2):176–187. https://doi.org/10.1093/ckj/sfw129

66. Colling R, Pitman H, Oien K, Rajpoot N, Macklin P, Snead D, Sackville T, Verrill C, Group CM-PAiHW (2019) Artificial intelligence in digital pathology: a roadmap to routine use in clinical practice. J Pathol. https://doi.org/10.1002/path.5310

67. Niazi MKK, Parwani AV, Gurcan MN (2019) Digital pathology and artificial intelligence. Lancet Oncol 20(5):e253–e261. https://doi.org/10.1016/S1470-2045(19)30154-8

Index

Rahul Vijay (ed.), *MERS Coronavirus: Methods and Protocols*, Methods in Molecular Biology, vol. 2099, https://doi.org/10.1007/978-1-0716-0211-9,

S

T

V

W

X

MIX
Papier aus verantwortungsvollen Quellen
Paper from responsible sources
FSC® C105338

Printed by Books on Demand, Germany